ESSAYS OF
SCHOPENHAUER

CONTENTS

PRELIMINARY.

When Schopenhauer was asked where he wished to be buried, he answered, "Anywhere; they will find me;" and the stone that marks his grave at Frankfort bears merely the inscription "Arthur Schopenhauer," without even the date of his birth or death. Schopenhauer, the pessimist, had a sufficiently optimistic conviction that his message to the world would ultimately be listened to—a conviction that never failed him during a lifetime of disappointments, of neglect in quarters where perhaps he would have most cherished appreciation; a conviction that only showed some signs of being justified a few years before his death. Schopenhauer was no opportunist; he was not even conciliatory; he never hesitated to declare his own faith in himself, in his principles, in his philosophy; he did not ask to be listened to as a matter of courtesy but as a right—a right for which he would struggle, for which he fought, and which has in the course of time, it may be admitted, been conceded to him.

Although everything that Schopenhauer wrote was written more or less as evidence to support his main philosophical thesis, his unifying philosophical principle, the essays in this volume have an interest, if not altogether apart, at least of a sufficiently independent interest to enable them to be considered on their own merits, without relation to his main idea. And in dissociating them, if one may do so for a moment (their author would have scarcely permitted it!), one feels that one enters a field of criticism in which opinions can scarcely vary. So far as his philosophy is concerned, this unanimity does not exist; he is one of the best abused amongst philosophers; he has many times been explained and condemned exhaustively, and no doubt this will be as many times repeated. What the trend of his underlying philosophical principal was, his metaphysical explanation of the world, is indicated in almost all the following essays, but chiefly in the "Metaphysics of Love," to which the reader may be referred.

These essays are a valuable criticism of life by a man who had a wide experience of life, a man of the world, who possessed an almost inspired faculty of observation. Schopenhauer, of all men, unmistakably observed life at first hand. There is no academic echo in his utterances; he is not one of a school; his voice has no formal intonation; it is deep, full-chested, and rings out its words with all the poignancy of individual emphasis, without bluster, but with unfailing conviction. He was for his time, and for his country, an adept at literary form; but he used it only as a means. Complicated as his sentences occasionally are, he says many sharp, many brilliant, many epigrammatic things, he has the manner of the famous essayists, he is paradoxical (how many of his paradoxes are now truisms!); one fancies at times that one is almost listening to a creation of Moliere, but these fireworks are not merely a literary display, they are used to illumine what he considers to be the truth. *Rien n'est beau que le vrai; le vrai seul est aimable,* he quotes; he was a deliberate and diligent searcher after truth, always striving to attain the heart of things, to arrive at a knowledge of first principles. It is, too, not without a sort of grim humour that this psychological vivisectionist attempts to lay bare the skeleton of the human mind, to tear away all the charming little sentiments and hypocrisies which in the course of time become a part and parcel of human life. A man influenced by such motives, and possessing a frank and caustic tongue, was not likely to attain any very large share of popular favour or to be esteemed a companionable sort of person. The fabric of social life is interwoven with a multitude of delicate evasions, of small hypocrisies, of matters of tinsel sentiment; social intercourse would be impossible, if it were not so. There is no sort of social existence possible for a person who is ingenuous enough to say always what he thinks, and, on the whole, one may be thankful that there is not. One naturally enough objects to form the subject of a critical diagnosis and exposure; one chooses for one's friends the agreeable hypocrites of life who sustain for one the illusions in which one wishes to live. The mere conception of a plain-speaking world is calculated to reduce one to the last degree of despair; it is the conception of the intolerable. Nevertheless it is good for mankind now and again to have a plain speaker, a "mar feast," on the scene; a wizard who devises for us a spectacle of disillusionment, and lets us for a moment see things as he honestly conceives them to be, and not as we would have them to be. But in estimating the value of a lesson of this sort, we must not be carried too far, not be altogether convinced. We may first take into account the temperament of the teacher; we may ask, is his vision perfect? We may indulge in a trifling diagnosis on our own account. And in an examination of this sort we find that Schopenhauer stands the test pretty well, if not with complete success. It strikes us that he suffers perhaps a little from a hereditary taint, for we know that there is an unmistakable predisposition to hypochondria in his family; we know, for instance, that his paternal grandmother became practically insane towards the end of her life, that two of her children suffered from some sort of mental incapacity, and that a third, Schopenhauer's father, was a man of curious temper and that he probably ended his own life. He himself would also have attached some importance, in a consideration of this sort, to the fact, as he might have put it, that his mother, when she married, acted in the interests of the individual instead of unconsciously fulfilling the will of the species, and that the offspring of the union suffered in consequence. Still, taking all these things into account, and attaching to them what importance they may be worth, one is amazed at the clearness of his vision, by his vigorous and at moments subtle perception. If he did not see life whole, what he did see he saw with his own eyes, and then told us all about it with unmistakable veracity, and for the most part simply, brilliantly. Too much importance cannot be attached to this quality of seeing things for oneself; it is the stamp of a great and original mind; it is the principal quality of what one calls genius.

In possessing Schopenhauer the world possesses a personality the richer; a somewhat garrulous personality it may be; a curiously whimsical and sensitive personality, full of quite ordinary superstitions, of extravagant vanities, selfish, at times violent, rarely generous; a man whom during his lifetime nobody quite knew, an isolated creature, self-absorbed, solely concerned in his elaboration of the explanation of the world, and possessing subtleties which for the most part escaped the perception of his fellows; at once a hermit and a boulevardier. His was essentially a great temperament; his whole life was a life of ideas, an intellectual life. And his work, the fruit of his life, would seem to be standing the test of all great work—the test of time. It is not a little curious that one so little realised in his own day, one so little lovable and so little loved, should now speak to us from his pages with something of the force of personal utterance, as if he were actually with us and as if we knew him, even as we know Charles Lamb and Izaak Walton, personalities of such a different calibre. And this man whom we realise does not impress us unfavourably; if he is without charm, he is surely immensely interesting and attractive; he is so strong in his intellectual convictions, he is so free from intellectual affectations, he is such an ingenuous egotist, so naively human; he is so mercilessly honest and independent, and, at times (one may be permitted to think), so mistaken.

R.D.

BIOGRAPHICAL NOTE.

Arthur Schopenhauer was born at No. 117 of the Heiligengeist Strasse, at Dantzic, on February 22, 1788. His parents on both sides traced their descent from Dutch ancestry, the great-grandfather of his mother having occupied some ecclesiastical position at Gorcum. Dr. Gwinner in his *Life* does not follow the Dutch ancestry on the father's side, but merely states that the great-grandfather of Schopenhauer at the beginning of the eighteenth century rented a farm, the Stuthof, in the neighbourhood of Dantzic. This ancestor, Andreas Schopenhauer, received here on one occasion an unexpected visit from Peter the Great and Catherine, and it is related that there being no stove in the chamber which the royal pair selected for the night, their host, for the purpose of heating it, set fire to several small bottles of brandy which had been emptied on the stone floor. His son Andreas followed in the footsteps of his father, combining a commercial career with country pursuits. He died in 1794 at Ohra, where he had purchased an estate, and to which he had retired to spend his closing years. His wife (the grandmother of Arthur) survived him for some years, although shortly after his death she was declared insane and incapable of managing her affairs. This couple had four sons: the eldest, Michael Andreas, was weak-minded; the second, Karl Gottfried, was also mentally weak and had deserted his people for evil companions; the youngest son, Heinrich Floris, possessed, however, in a considerable degree the qualities which his brothers lacked. He possessed intelligence, a strong character, and had great commercial sagacity; at the same time, he took a definite interest in intellectual pursuits, reading Voltaire, of whom he was more or less a disciple, and other French authors, possessing a keen admiration for English political and family life, and furnishing his house after an English fashion. He was a man of fiery temperament and his appearance was scarcely prepossessing; he was short and stout; he had a broad face and turned-up nose, and a large mouth. This was the father of our philosopher.

When he was thirty-eight, Heinrich Schopenhauer married, on May 16, 1785, Johanna Henriette Trosiener, a young lady of eighteen, and daughter of a member of the City Council of Dantzic. She was at this time an attractive, cultivated young person, of a placid disposition, who seems to have married more because marriage offered her a comfortable settlement and assured position in life, than from any passionate affection for her wooer, which, it is just to her to say, she did not profess. Heinrich Schopenhauer was so much influenced by English ideas that he desired that his first child should be born in England; and thither, some two years after their marriage, the pair, after making a *detour* on the Continent, arrived. But after spending some weeks in London Mrs. Schopenhauer was seized with home-sickness, and her husband acceded to her entreaties to return to Dantzic, where a child, the future philosopher, was shortly afterwards born. The first five years of the child's life were spent in the country, partly at the Stuthof which had formerly belonged to Andreas Schopenhauer, but had recently come into the possession of his maternal grandfather.

Five years after the birth of his son, Heinrich Schopenhauer, in consequence of the political crisis, which he seems to have taken keenly to heart, in the affairs of the Hanseatic town of Dantzic, transferred his business and his home to Hamburg, where in 1795 a second child, Adele, was born. Two years later, Heinrich, who intended to train his son for a business life, took him, with this idea, to Havre, by way of Paris, where they spent a little time, and left him there with M. Gregoire, a commercial connection. Arthur remained at Havre for two years, receiving private instruction with this man's son Anthime, with whom he struck up a strong friendship, and when he returned to Hamburg it was

14

found that he remembered but few words of his mother-tongue. Here he was placed in one of the principal private schools, where he remained for three years. Both his parents, but especially his mother, cultivated at this time the society of literary people, and entertained at their house Klopstock and other notable persons. In the summer following his return home from Havre he accompanied his parents on a continental tour, stopping amongst other places at Weimar, where he saw Schiller. His mother, too, had considerable literary tastes, and a distinct literary gift which, later, she cultivated to some advantage, and which brought her in the production of accounts of travel and fiction a not inconsiderable reputation. It is, therefore, not surprising that literary tendencies began to show themselves in her son, accompanied by a growing distaste for the career of commerce which his father wished him to follow. Heinrich Schopenhauer, although deprecating these tendencies, considered the question of purchasing a canonry for his son, but ultimately gave up the idea on the score of expense. He then proposed to take him on an extended trip to France, where he might meet his young friend Anthime, and then to England, if he would give up the idea of a literary calling, and the proposal was accepted.

In the spring of 1803, then, he accompanied his parents to London, where, after spending some time in sight-seeing, he was placed in the school of Mr. Lancaster at Wimbledon. Here he remained for three months, from July to September, laying the foundation of his knowledge of the English language, while his parents proceeded to Scotland. English formality, and what he conceived to be English hypocrisy, did not contrast favourably with his earlier and gayer experiences in France, and made an extremely unfavourable impression upon his mind; which found expression in letters to his friends and to his mother.

On returning to Hamburg after this extended excursion abroad, Schopenhauer was placed in the office of a Hamburg senator called Jenisch, but he was as little inclined as ever to follow a commercial career, and secretly shirked his work so that he might pursue his studies. A little later a somewhat unexplainable calamity occurred. When Dantzic ceased to be a free city, and Heinrich Schopenhauer at a considerable cost and monetary sacrifice transferred his business to Hamburg, the event caused him much bitterness of spirit. At Hamburg his business seems to have undergone fluctuations. Whether these further affected his spirit is not sufficiently established, but it is certain, however, that he developed peculiarities of manner, and that his temper became more violent. At any rate, one day in April 1805 it was found that he had either fallen or thrown himself into the canal from an upper storey of a granary; it was generally concluded that it was a case of suicide.

Schopenhauer was seventeen at the time of this catastrophe, by which he was naturally greatly affected. Although by the death of his father the influence which impelled him to a commercial career was removed, his veneration for the dead man remained with him through life, and on one occasion found expression in a curious tribute to his memory in a dedication (which was not, however, printed) to the second edition of *Die Welt als Wille und Vorstellung.* "That I could make use of and cultivate in a right direction the powers which nature gave me," he concludes, "that I could follow my natural impulse and think and work for countless others without the help of any one; for that I thank thee, my father, thank thy activity, thy cleverness, thy thrift and care for the future. Therefore I praise thee, my noble father. And every one who from my work derives any pleasure, consolation, or instruction shall hear thy name and know that if Heinrich Floris Schopenhauer had not been the man he was, Arthur Schopenhauer would have been a hundred times ruined."

The year succeeding her husband's death, Johanna Schopenhauer removed with her daughter to Weimar, after having attended to the settlement of her husband's affairs, which left her in possession of a considerable income. At Weimar she devoted herself to the pursuit of literature, and held twice a week a sort of salon, which was attended by Goethe, the two Schlegels, Wieland, Heinrich Meyer, Grimm, and other literary persons of note. Her son meanwhile continued for another year at the "dead timber of the desk," when his mother, acting under the advice of her friend Fernow, consented, to his great joy, to his following his literary bent.

During the next few years we find Schopenhauer devoting himself assiduously to acquiring the equipment for a learned career; at first at the Gymnasium at Gotha, where he penned some satirical verses on one of the masters, which brought him into some trouble. He removed in consequence to Weimar, where he pursued his classical studies under the direction of Franz Passow, at whose house he lodged. Unhappily, during his sojourn at Weimar his relations with his mother became strained. One feels that there is a sort of autobiographical interest in his essay on women, that his view was largely influenced by his relations with his mother, just as one feels that his particular argument in his essay on education is largely influenced by the course of his own training.

On his coming of age Schopenhauer was entitled to a share of the paternal estate, a share which yielded him a yearly income of about L150. He now entered himself at the University of Goettingen (October 1809), enrolling himself as a student of medicine, and devoting himself to the study of the natural sciences, mineralogy, anatomy, mathematics, and history; later, he included logic, physiology, and ethnography. He had always been passionately devoted to music and found relaxation in learning to play the flute and guitar. His studies at this time did not preoccupy him to the extent of isolation; he mixed freely with his fellows, and reckoned amongst his friends or acquaintances, F.W. Kreise, Bunsen, and Ernst Schulze. During one vacation he went on an expedition to Cassel and to the Hartz Mountains. It was about this time, and partly owing to the influence of Schulze, the author of *Aenesidemus*, and then a professor at the University of Goettingen, that Schopenhauer came to realise his vocation as that of a philosopher.

During his holiday at Weimar he called upon Wieland, then seventy-eight years old, who, probably prompted by Mrs. Schopenhauer, tried to dissuade him from the vocation which he had chosen. Schopenhauer in reply said, "Life is a difficult question; I have decided to spend my life in thinking about it." Then, after the conversation had continued for some little time, Wieland declared warmly that he thought that he had chosen rightly. "I understand your nature," he said; "keep to philosophy." And, later, he told Johanna Schopenhauer that he thought her son would be a great man some day.

Towards the close of the summer of 1811 Schopenhauer removed to Berlin and entered the University. He here continued his study of the natural sciences; he also attended the lectures on the History of Philosophy by Schleiermacher, and on Greek Literature and Antiquities by F.A. Wolf, and the lectures on "Facts of Consciousness" and "Theory of Science" by Fichte, for the last of whom, as we know indeed from frequent references in his books, he had no little contempt. A year or so later, when the news of Napoleon's disaster in Russia arrived, the Germans were thrown into a state of great excitement, and made speedy preparations for war. Schopenhauer contributed towards equipping volunteers for the army, but he did not enter active service; indeed, when the result of the battle of Luetzen was known and Berlin seemed to be in danger, he fled for safety to Dresden and thence to Weimar. A little later we find him at Rudolstadt, whither he had proceeded in consequence of the recurrence of differences with his mother, and remained there from June to November 1813, principally engaged in the composition of an essay, "A Philosophical Treatise on the Fourfold Root of the Principle of Sufficient Reason," which he offered to the University of Jena as an exercise to qualify for the degree of Doctor of Philosophy, and for which a diploma was granted. He published this essay at his own cost towards the end of the year, but it seems to have fallen flatly from the press, although its arguments attracted the attention and the sympathy of Goethe, who, meeting him on his return to Weimar in November, discussed with him his own theory of colour. A couple of years before, Goethe, who was opposed to the Newtonian theory of light, had brought out his *Farbenlehre* (colour theory). In Goethe's diary Schopenhauer's name frequently occurs, and on the 24th November 1813 he wrote to Knebel: "Young Schopenhauer is a remarkable and interesting man . . . I find him intellectual, but I am undecided about him as far as other things go." The result of this association with Goethe was his *Ueber das Sehn und die Farben* ("On Vision and Colour"), published at Leipzig in 1816, a copy of which he forwarded to Goethe (who had already seen the MS.) on the 4th May of that year. A few days later Goethe wrote to the distinguished scientist, Dr. Seebeck, asking him to read the work. In Gwinner's *Life* we find the copy of a letter written in English to Sir C.L. Eastlake: "In the year 1830, as I was going to publish in Latin the same treatise which in German accompanies this letter, I went to Dr. Seebeck of the Berlin Academy, who is universally admitted to be the first natural philosopher (in the English sense of the word meaning physiker) of Germany; he is the discoverer of thermo-electricity and of several physical truths. I questioned him on his opinion on the controversy between Goethe and Newton; he was extremely cautious and made me promise that I should not print and publish anything of what he might say, and at last, being hard pressed by me, he confessed that indeed Goethe was perfectly right and Newton wrong, but that he had no business to tell the world so. He has died since, the old coward!"

In May 1814 Schopenhauer removed from Weimar to Dresden, in consequence of the recurrence of domestic differences with his mother. This was the final break between the pair, and he did not see her again during the remaining twenty-four years of her life, although they resumed correspondence some years before her death. It were futile to attempt to revive the dead bones of the cause of these unfortunate differences between Johanna Schopenhauer and her son. It was a question of opposing temperaments; both and neither were at once to blame. There is no reason to suppose that Schopenhauer was ever a conciliatory son, or a companionable person to live with; in fact, there is plenty to show that he possessed trying and irritating qualities, and that he assumed an attitude of criticism towards his mother that could not in any circumstances be agreeable. On the other hand, Anselm Feuerbach in his *Memoirs* furnishes us with a scarcely prepossessing picture of Mrs. Schopenhauer: "Madame Schopenhauer," he writes, "a rich widow. Makes profession of erudition. Authoress. Prattles much and well, intelligently; without heart and soul. Self-complacent, eager after approbation, and constantly smiling to herself. God preserve us from women whose mind has shot up into mere intellect."

Schopenhauer meanwhile was working out his philosophical system, the idea of his principal philosophical work. "Under my hands," he wrote in 1813, "and still more in my mind grows a work, a philosophy which will be an ethics and a metaphysics in one:—two branches which hitherto have been separated as falsely as man has been divided into soul and body. The work grows, slowly and gradually aggregating its parts like the child in the womb. I became aware of one member, one vessel, one part after another. In other words, I set each sentence down without anxiety as to how it will fit into the whole; for I know it has all sprung from a single foundation. It is thus that an organic whole originates, and that alone will live . . . Chance, thou ruler of this sense-world! Let me live and find peace for yet a few years, for I love my work as the mother her child. When it is matured and has come to birth, then exact from me thy duties, taking interest for the postponement. But, if I sink before the time in this iron age, then grant that these miniature beginnings, these studies of mine, be given to the world as they are and for what they are: some day perchance will arise a kindred spirit, who can frame the members together and 'restore' the fragment of antiquity."[1]

By March 1817 he had completed the preparatory work of his system, and began to put the whole thing together; a year later *Die Welt als Wille und Vorstellung: vier Buecher, nebst einem Anhange, der die Kritik der Kantischen Philosophie enthaelt* ("The World as Will and Idea; four books, with an appendix containing a criticism on the philosophy of Kant"). Some delay occurring in the publication, Schopenhauer wrote one of his characteristically abusive letters to Brockhaus, his publisher, who retorted "that he must decline all further correspondence with one whose letters, in their divine coarseness and rusticity, savoured more of the cabman than of the philosopher," and concluded with a hope that his fears that the work he was printing would be good for nothing but waste paper, might not be realised.[2] The work appeared about the end of December 1818 with 1819 on the title-page. Schopenhauer had meanwhile proceeded in September to Italy, where he revised the final proofs. So far as the reception of the work was concerned there was reason to believe that the fears of Brockhaus would be realised, as, in fact, they came practically to be. But in the face of this general want of appreciation, Schopenhauer had some crumbs of consolation. His sister wrote to him in March (he was then staying at Naples) that Goethe "had received it with great joy, immediately cut the thick book, and began *instantly* to read it. An hour later he sent me a note to say that he thanked you very much and thought that the whole book was good. He pointed out the most important passages, read them to us, and was greatly delighted . . . You are the only author whom Goethe has ever read seriously, it seems to me, and I rejoice." Nevertheless the book did not sell. Sixteen years later Brockhaus informed Schopenhauer that a large number of copies had been sold at waste paper price, and that he had even then a few in stock. Still, during the years 1842-43, Schopenhauer was contemplating the issue of a second edition and making revisions for that purpose; when he had completed the work he took it to Brockhaus, and agreed to leave the question of remuneration open. In the following year the second edition was issued (500 copies of the first volume, and 750 of the second), and for this the author was to receive no remuneration. "Not to my contemporaries," says Schopenhauer with fine conviction in his preface to this edition, "not to my compatriots—to mankind I commit my now completed work, in the confidence that it will not be without value for them, even if this should be late recognised, as is commonly the lot of what is good. For it cannot have been for the passing generation, engrossed with the delusion of the moment, that my mind, almost against my will, has uninterruptedly stuck to its work through the course of a long life. And while the lapse of time has not been able to make me doubt the worth of my work, neither has the lack of sympathy; for I constantly saw the false and the bad, and finally the absurd and senseless, stand in universal admiration and honour, and I bethought myself that if it were not the case, those who are capable of recognising the genuine and right are so rare that we may look for them in vain for some twenty years, then those who are capable of producing it could not be so few that their works afterwards form an exception to the perishableness of earthly things; and thus would be lost the reviving prospect of posterity which every one who sets before himself a high aim requires to strengthen him."[3]

When Schopenhauer started for Italy Goethe had provided him with a letter of introduction to Lord Byron, who was then staying at Venice, but Schopenhauer never made use of the letter; he said that he hadn't the courage to present himself. "Do you know," he says in a letter, "three great pessimists were in Italy at the same time—Byron, Leopardi, and myself! And yet not one of us has made the acquaintance of the other." He remained in Italy until June 1819, when he proceeded to Milan, where he received distressing news from his sister to the effect that a Dantzic firm, in which she and her mother had invested all their capital, and in which he himself had invested a little, had become bankrupt. Schopenhauer immediately proposed to share his own income with them. But later, when the defaulting firm offered to its creditors a composition of thirty per cent, Schopenhauer would accept nothing less than seventy per cent in the case of immediate payment, or the whole if the payment were deferred; and he was so indignant at his mother and sister falling in with the arrangement of the debtors, that he did not correspond with them again for eleven years. With reference to this affair he wrote: "I can imagine that from your point of view my behaviour may seem hard and unfair. That is a mere illusion which disappears as soon as you reflect that all I want is merely not to have taken from me what is most rightly and incontestably mine, what, moreover, my whole happiness, my freedom, my learned leisure depend upon;—a blessing which in this world people like me enjoy so rarely that it would be almost as unconscientious as cowardly not to defend it to the uttermost and maintain it by every exertion. You say, perhaps, that if all your creditors were of this way of thinking, I too should come badly off. But if all men thought as I do, there would be much more thinking done, and in that case probably there would be neither bankruptcies, nor wars, nor gaming tables."[4]

In July 1819, when he was at Heidelberg, the idea occurred to him of turning university lecturer, and took practical shape the following summer, when he delivered a course of lectures on philosophy at the Berlin University. But the experiment was not a success; the course was not completed through the want of attendance, while Hegel at the same time and place was lecturing to a crowded and enthusiastic audience. This failure embittered him, and during the next few years there is little of any moment in his life to record. There was one incident, however, to which his detractors would seem to have attached more importance than it was worth, but which must have been sufficiently disturbing to Schopenhauer—we refer to the Marquet affair. It appears on his returning home one day he found three women gossiping outside his door, one of whom was a seamstress who occupied another room in the house. Their presence irritated Schopenhauer (whose sensitiveness in such matters may be estimated from his essay "On Noise"), who, finding them occupying the same position on another occasion, requested them to go away, but the seamstress replied that she was an honest person and refused to move. Schopenhauer disappeared into his apartments and returned with a stick. According to his own account, he offered his arm to the woman in order to take her out; but she would not accept it, and remained where she was. He then threatened to put her out, and carried his threat into execution by seizing her round the waist and putting her out. She screamed, and attempted to return. Schopenhauer now pushed her out; the woman fell, and raised the whole house. This woman, Caroline Luise Marquet, brought an action against him for damages, alleging that he had kicked

23

and beaten her. Schopenhauer defended his own case, with the result that the action was dismissed. The woman appealed, and Schopenhauer, who was contemplating going to Switzerland, did not alter his plans, so that the appeal was heard during his absence, the judgment reversed, and he was mulcted in a fine of twenty thalers. But the unfortunate business did not end here. Schopenhauer proceeded from Switzerland to Italy, and did not return to Berlin until May 1825. Caroline Marquet renewed her complaints before the courts, stating that his ill-usage had occasioned a fever through which she had lost the power of one of her arms, that her whole system was entirely shaken, and demanding a monthly allowance as compensation. She won her case; the defendant had to pay three hundred thalers in costs and contribute sixty thalers a year to her maintenance while she lived. Schopenhauer on returning to Berlin did what he could to get the judgment reversed, but unsuccessfully. The woman lived for twenty years; he inscribed on her death certificate, *"Obit anus, obit onus"*

The idea of marriage seems to have more or less possessed Schopenhauer about this time, but he could not finally determine to take the step. There is sufficient to show in the following essays in what light he regarded women. Marriage was a debt, he said, contracted in youth and paid off in old age. Married people have the whole burden of life to bear, while the unmarried have only half, was a characteristically selfish apothegm. Had not all the true philosophers been celibates—Descartes, Leibnitz, Malebranche, Spinoza, and Kant? The classic writers were of course not to be considered, because with them woman occupied a subordinate position. Had not all the great poets married, and with disastrous consequences? Plainly, Schopenhauer was not the person to sacrifice the individual to the will of the species.

In August 1831 he made a fortuitous expedition to Frankfort-on-the-Main—an expedition partly prompted by the outbreak of cholera at Berlin at the time, and partly by the portent of a dream (he was credulous in such matters) which at the beginning of the year had intimated his death. Here, however, he practically remained until his death, leading

24

a quiet, mechanically regular life and devoting his thoughts to the development of his philosophic ideas, isolated at first, but as time went on enjoying somewhat greedily the success which had been denied him in his earlier days. In February 1839 he had a moment of elation when he heard from the Scientific Society of Drontheim that he had won the prize for the best essay on the question, "Whether free will could be proved from the evidence of consciousness," and that he had been elected a member of the Society; and a corresponding moment of despondency when he was informed by the Royal Danish Academy of the Sciences at Copenhagen, in a similar competition, that his essay on "Whether the source and foundation of ethics was to be sought in an intuitive moral idea, and in the analysis of other derivative moral conceptions, or in some other principle of knowledge," had failed, partly on the ground of the want of respect which it showed to the opinions of the chief philosophers. He published these essays in 1841 under the title of "The Two Fundamental Problems of Ethics," and ten years later *Parerga und Paralipomena* the composition of which had engaged his attention for five or six years. The latter work, which proved to be his most popular, was refused by three publishers, and when eventually it was accepted by Hayn of Berlin, the author only received ten free copies of his work as payment. It is from this book that all except one of the following essays have been selected; the exception is "The Metaphysics of Love," which appears in the supplement of the third book of his principal work. The second edition of *Die Welt als Wille und Vorstellung* appeared in 1844, and was received with growing appreciation. Hitherto he had been chiefly known in Frankfort as the son of the celebrated Johanna Schopenhauer; now he came to have a following which, if at first small in numbers, were sufficiently enthusiastic, and proved, indeed, so far as his reputation was concerned, helpful. Artists painted his portrait; a bust of him was made by Elizabeth Ney. In the April number of the *Westminster Review* for 1853 John Oxenford, in an article entitled "Iconoclasm in German Philosophy," heralded in England his recognition as a writer and thinker;

three years later Saint-Rene Taillandier, in the *Revue des Deux Mondes*, did a similar service for him in France. One of his most enthusiastic admirers was Richard Wagner, who in 1854 sent him a copy of his *Der Ring der Nibelungen*, with the inscription "In admiration and gratitude." The Philosophical Faculty of the University of Leipzic offered a prize for an exposition and criticism of his philosophical system. Two Frenchmen, M. Foucher de Careil and M. Challemel Lacour, who visited Schopenhauer during his last days, have given an account of their impressions of the interview, the latter in an article entitled, "Un Bouddhiste Contemporain en Allemagne," which appeared in the *Revue des Deux Mondes* for March 15th, 1870. M. Foucher de Careil gives a charming picture of him:—

"Quand je le vis, pour la premiere fois, en 1859, a la table de l'hotel d'Angleterre, a Francfort, c'etait deja un vieillard, a l'oeil d'un bleu vif et limpide, a la levre mince et legerement sarcastique, autour de laquelle errait un fin sourire, et dont le vaste front, estompe de deux touffes de cheveux blancs sur les cotes, relevait d'un cachet de noblesse et de distinction la physionomie petillante d'esprit et de malice. Les habits, son jabot de dentelle, sa cravate blanche rappelaient un vieillard de la fin du regne de Louis XV; ses manieres etaient celles d'un homme de bonne compagnie. Habituellement reserve et d'un naturel craintif jusqu'a la mefiance, il ne se livrait qu'avec ses intimes ou les etrangers de passage a Francfort. Ses mouvements etaient vifs et devenaient d'une petulance extraordinaire dans la conversation; il fuyait les discussions et les vains combats de paroles, mais c'etait pour mieux jouir du charme d'une causerie intime. Il possedait et parlait avec une egale perfection quatre langues: le francais, l'anglais, l'allemand, l'italien et passablement l'espagnol. Quand il causait, la verve du vieillard brodait sur le canevas un peu lourd de l'allemand ses brilliantes arabesques latines, grecques, francaises, anglaises, italiennes. C'etait un entrain, une precision et des sailles, une richesse de citations, une exactitude de details qui faisait couler les heures; et quelquefois le petit cercle de ses intimes l'ecoutait jusqu'a minuit, sans qu'un moment de fatigue se fut peint sur ses traits ou que le feu de son regard se fut un instant amorti. Sa parole nette et accentuee captivait l'auditoire: elle peignait et analysait tout ensemble; une sensibilite delicate en augmentait le feu; elle etait exacte et precise sur toutes sortes de sujets."

Schopenhauer died on the 20th September 1860, in his seventy-third year, peacefully, alone as he had lived, but not without warning. One day in April, taking his usual brisk walk after dinner, he suffered from palpitation of the heart, he could scarcely breathe. These symptoms developed during the next few months, and Dr. Gwinner advised him to discontinue his cold baths and to breakfast in bed; but Schopenhauer, notwithstanding his early medical training, was little inclined to follow medical advice. To Dr. Gwinner, on the evening of the 18th September, when he expressed a hope that he might be able to go to Italy, he said that it would be a pity if he died now, as he wished to make several important additions to his *Parerga*; he spoke about his works and of the warm recognition with which they had been welcomed in the most remote places. Dr. Gwinner had never before found him so eager and gentle, and left him reluctantly, without, however, the least premonition that he had seen him for the last time. On the second morning after this interview Schopenhauer got up as usual, and had his cold bath and breakfast. His servant had opened the window to let in the morning air and had then left him. A little later Dr. Gwinner arrived and found him reclining in a corner of the sofa; his face wore its customary expression; there was no sign of there having been any struggle with death. There had been no struggle with death; he had died, as he had hoped he would die, painlessly, easily.

In preparing the above notice the writer has to acknowledge her indebtedness to Dr. Gwinner's *Life* and Professor Wallace's little work on the same subject, as well as to the few other authorities that have been available.—THE TRANSLATOR.

FOOTNOTES:

1   Wallace's *Life*, pp. 95, 96.
2   Wallace, p. 108.
3   Haldane and Kemp's *The World as Will and Idea.*
4   Wallace, p. 145.

ON AUTHORSHIP AND STYLE.

There are, first of all, two kinds of authors: those who write for the subject's sake, and those who write for writing's sake. The first kind have had thoughts or experiences which seem to them worth communicating, while the second kind need money and consequently write for money. They think in order to write, and they may be recognised by their spinning out their thoughts to the greatest possible length, and also by the way they work out their thoughts, which are half-true, perverse, forced, and vacillating; then also by their love of evasion, so that they may seem what they are not; and this is why their writing is lacking in definiteness and clearness.

Consequently, it is soon recognised that they write for the sake of filling up the paper, and this is the case sometimes with the best authors; for example, in parts of Lessing's *Dramaturgie,* and even in many of Jean Paul's romances. As soon as this is perceived the book should be thrown away, for time is precious. As a matter of fact, the author is cheating the reader as soon as he writes for the sake of filling up paper; because his pretext for writing is that he has something to impart. Writing for money and preservation of copyright are, at bottom, the ruin of literature. It is only the man who writes absolutely for the sake of the subject that writes anything worth writing. What an inestimable advantage it would be, if, in every branch of literature, there existed only a few but excellent books! This can never come to pass so long as money is to be made by writing. It seems as if money lay under a curse, for every author deteriorates directly he writes in any way for the sake of money. The best works of great men all come from the time when they had to write either for nothing or for very little pay. This is confirmed by the Spanish proverb: *honra y provecho no caben en un saco* (Honour and money are not to be found in the same purse). The deplorable condition of the literature of to-day, both in Germany and other countries, is due to the fact that books are written for the sake of earning money. Every one who is in want of money sits down and writes a book, and the public is stupid enough to buy it. The secondary effect of this is the ruin of language.

A great number of bad authors eke out their existence entirely by the foolishness of the public, which only will read what has just been printed. I refer to journalists, who have been appropriately so-called. In other words, it would be "day labourer."

* * * * *

Again, it may be said that there are three kinds of authors. In the first place, there are those who write without thinking. They write from memory, from reminiscences, or even direct from other people's books. This class is the most numerous. In the second, those who think whilst they are writing. They think in order to write; and they are numerous. In the third place, there are those who have thought before they begin to write. They write solely because they have thought; and they are rare.

Authors of the second class, who postpone their thinking until they begin to write, are like a sportsman who goes out at random—he is not likely to bring home very much. While the writing of an author of the third, the rare class, is like a chase where the game has been captured beforehand and cooped up in some enclosure from which it is afterwards set free, so many at a time, into another enclosure, where it is not possible for it to escape, and the sportsman has now nothing to do but to aim and fire—that is to say, put his thoughts on paper. This is the kind of sport which yields something.

But although the number of those authors who really and seriously think before they write is small, only extremely few of them think about *the subject itself*; the rest think only about the books written on this subject, and what has been said by others upon it, I mean. In order to think, they must have the more direct and powerful incentive of other people's thoughts. These become their next theme, and therefore they always remain under their influence and are never, strictly speaking, original. On the contrary, the former are roused to thought through the *subject itself*, hence their thinking is directed immediately to it. It is only among them that we find the authors whose names become immortal. Let it be understood that I am speaking here of writers of the higher branches of literature, and not of writers on the method of distilling brandy.

It is only the writer who takes the material on which he writes direct out of his own head that is worth reading. Book manufacturers, compilers, and the ordinary history writers, and others like them, take their material straight out of books; it passes into their fingers without its having paid transit duty or undergone inspection when it was in their heads, to say nothing of elaboration. (How learned many a man would be if he knew everything that was in his own books!) Hence their talk is often of such a vague nature that one racks one's brains in vain to understand of *what* they are really thinking. They are not thinking at all. The book from which they copy is sometimes composed in the same way: so that writing of this kind is like a plaster cast of a cast of a cast, and so on, until finally all that is left is a scarcely recognisable outline of the face of Antinous. Therefore, compilations should be read as seldom as possible: it is difficult to avoid them entirely, since compendia, which contain in a small space knowledge that has been collected in the course of several centuries, are included in compilations.

No greater mistake can be made than to imagine that what has been written latest is always the more correct; that what is written later on is an improvement on what was written previously; and that every change means progress. Men who think and have correct judgment, and people who treat their subject earnestly, are all exceptions only. Vermin is the rule everywhere in the world: it is always at hand and busily engaged in trying to improve in its own way upon the mature deliberations of the thinkers. So that if a man wishes to improve himself in any subject he must guard against immediately seizing the newest books written upon it, in the assumption that science is always advancing and that the older books have been made use of in the compiling of the new. They have, it is true, been used; but how? The writer often does not thoroughly understand the old books; he will, at the same time, not use their exact words, so that the result is he spoils and bungles what has been said in a much better and clearer way by the old writers; since they wrote from their own lively knowledge of the subject. He often leaves out the best things they have written, their most striking elucidations of the matter, their happiest remarks, because he does not recognise their value or feel how pregnant they are. It is only what is stupid and shallow that appeals to him. An old and excellent book is frequently shelved for new and bad ones; which, written for the sake of money, wear a pretentious air and are much eulogised by the authors' friends. In science, a man who wishes to distinguish himself brings something new to market; this frequently consists in his denouncing some principle that has been previously held as correct, so that he may establish a wrong one of his own. Sometimes his attempt is successful for a short time, when a return is made to the old and correct doctrine. These innovators are serious about nothing else in the world than their own priceless person, and it is this that they wish to make its mark. They bring this quickly about by beginning a paradox; the sterility of their own heads suggests their taking the path of negation; and truths that have long been recognised are now denied— for instance, the vital power, the sympathetic nervous system, *generatio equivoca*, Bichat's distinction between the working of the passions and the working of intelligence, or they return to crass atomism, etc., etc. Hence *the course of science is often retrogressive.*

To this class of writers belong also those translators who, besides translating their author, at the same time correct and alter him, a thing that always seems to me impertinent. Write books yourself which are

worth translating and leave the books of other people as they are. One
should read, if it is possible, the real authors, the founders and discoverers
of things, or at any rate the recognised great masters in every branch of
learning, and buy second-hand *books* rather than read their *contents* in new
ones.

It is true that *inventis aliquid addere facile est*, therefore a man, after
having studied the principles of his subject, will have to make himself
acquainted with the more recent information written upon it. In general,
the following rule holds good here as elsewhere, namely: what is new is
seldom good; because a good thing is only new for a short time.

What the address is to a letter the *title* should be to a book—that is, its
immediate aim should be to bring the book to that part of the public that
will be interested in its contents. Therefore, the title should be effective,
and since it is essentially short, it should be concise, laconic, pregnant,
and if possible express the contents in a word. Therefore a title that is
prolix, or means nothing at all, or that is indirect or ambiguous, is bad;
so is one that is false and misleading: this last may prepare for the book
the same fate as that which awaits a wrongly addressed letter. The worst
titles are those that are stolen, such titles that is to say that other books
already bear; for in the first place they are a plagiarism, and in the second
a most convincing proof of an absolute want of originality. A man who
has not enough originality to think out a new title for his book will be
much less capable of giving it new contents. Akin to these are those titles
which have been imitated, in other words, half stolen; for instance, a long
time after I had written "On Will in Nature," Oersted wrote "On Mind
in Nature."

* * * * *

A book can never be anything more than the impression of its author's
thoughts. The value of these thoughts lies either in the *matter about which*
he has thought, or in the *form* in which he develops his matter—that is to
say, *what* he has thought about it.

33

The matter of books is very various, as also are the merits conferred on books on account of their matter. All matter that is the outcome of experience, in other words everything that is founded on fact, whether it be historical or physical, taken by itself and in its widest sense, is included in the term matter. It is the *motif* that gives its peculiar character to the book, so that a book can be important whoever the author may have been; while with form the peculiar character of a book rests with the author of it. The subjects may be of such a nature as to be accessible and well known to everybody; but the form in which they are expounded, *what* has been thought about them, gives the book its value, and this depends upon the author. Therefore if a book, from this point of view, is excellent and without a rival, so also is its author. From this it follows that the merit of a writer worth reading is all the greater the less he is dependent on matter—and the better known and worn out this matter, the greater will be his merit. The three great Grecian tragedians, for instance, all worked at the same subject.

So that when a book becomes famous one should carefully distinguish whether it is so on account of its matter or its form.

Quite ordinary and shallow men are able to produce books of very great importance because of their *matter*, which was accessible to them alone. Take, for instance, books which give descriptions of foreign countries, rare natural phenomena, experiments that have been made, historical events of which they were witnesses, or have spent both time and trouble in inquiring into and specially studying the authorities for them.

On the other hand, it is on *form* that we are dependent, where the matter is accessible to every one or very well known; and it is what has been thought about the matter that will give any value to the achievement; it will only be an eminent man who will be able to write anything that is worth reading. For the others will only think what is possible for every other man to think. They give the impress of their own mind; but every one already possesses the original of this impression.

However, the public is very much more interested in matter than in form, and it is for this very reason that it is behindhand in any high degree of culture. It is most laughable the way the public reveals its liking for matter in poetic works; it carefully investigates the real events or personal circumstances of the poet's life which served to give the *motif* of his works; nay, finally, it finds these more interesting than the works themselves; it reads more about Goethe than what has been written by Goethe, and industriously studies the legend of Faust in preference to Goethe's *Faust* itself. And when Buerger said that "people would make learned expositions as to who Leonora really was," we see this literally fulfilled in Goethe's case, for we now have many learned expositions on Faust and the Faust legend. They are and will remain of a purely material character. This preference for matter to form is the same as a man ignoring the shape and painting of a fine Etruscan vase in order to make a chemical examination of the clay and colours of which it is made. The attempt to be effective by means of the matter used, thereby ministering to this evil propensity of the public, is absolutely to be censured in branches of writing where the merit must lie expressly in the form; as, for instance, in poetical writing. However, there are numerous bad dramatic authors striving to fill the theatre by means of the matter they are treating. For instance, they place on the stage any kind of celebrated man, however stripped of dramatic incidents his life may have been, nay, sometimes without waiting until the persons who appear with him are dead.

The distinction between matter and form, of which I am here speaking, is true also in regard to conversation. It is chiefly intelligence, judgment, wit, and vivacity that enable a man to converse; they give form to the conversation. However, the *matter* of the conversation must soon come into notice—in other words, *that* about which one can talk to the man, namely, his knowledge. If this is very small, it will only be his possessing the above-named formal qualities in a quite exceptionally high degree that will make his conversation of any value, for his matter will be restricted to things concerning humanity and nature, which are known generally. It is just the reverse if a man is wanting in these formal qualities, but has, on the other hand, knowledge of such a kind that it lends value to his conversation; this value, however, will then entirely rest on the matter of his conversation, for, according to the Spanish proverb, *mas sabe el necio en su casa, que el sabio en la agena.*

A thought only really lives until it has reached the boundary line of words; it then becomes petrified and dies immediately; yet it is as everlasting as the fossilised animals and plants of former ages. Its existence, which is really momentary, may be compared to a crystal the instant it becomes crystallised.

As soon as a thought has found words it no longer exists in us or is serious in its deepest sense.

When it begins to exist for others it ceases to live in us; just as a child frees itself from its mother when it comes into existence. The poet has also said:

"Ihr muesst mich nicht durch Widerspruch verwirren!
Sobald man spricht, beginnt man schon zu irren."

The pen is to thought what the stick is to walking, but one walks most easily without a stick, and thinks most perfectly when no pen is at hand. It is only when a man begins to get old that he likes to make use of a stick and his pen.

A hypothesis that has once gained a position in the mind, or been born in it, leads a life resembling that of an organism, in so far as it receives from the outer world matter only that is advantageous and homogeneous to it; on the other hand, matter that is harmful and heterogeneous to it is either rejected, or if it must be received, cast off again entirely.

Abstract and indefinite terms should be employed in satire only as they are in algebra, in place of concrete and specified quantities. Moreover, it should be used as sparingly as the dissecting knife on the body of a living man. At the risk of forfeiting his life it is an unsafe experiment.

For a work to become *immortal* it must possess so many excellences that it will not be easy to find a man who understands and values them *all*; so that there will be in all ages men who recognise and appreciate some of these excellences; by this means the credit of the work will be retained throughout the long course of centuries and ever-changing interests, for, as it is appreciated first in this sense, then in that, the interest is never exhausted.

An author like this, in other words, an author who has a claim to live on in posterity, can only be a man who seeks in vain his like among his contemporaries over the wide world, his marked distinction making him a striking contrast to every one else. Even if he existed through several generations, like the wandering Jew, he would still occupy the same position; in short, he would be, as Ariosto has put it, *lo fece natura, e poi ruppe lo stampo*. If this were not so, one would not be able to understand why his thoughts should not perish like those of other men.

In almost every age, whether it be in literature or art, we find that if a thoroughly wrong idea, or a fashion, or a manner is in vogue, it is admired. Those of ordinary intelligence trouble themselves inordinately to acquire it and put it in practice. An intelligent man sees through it and despises it, consequently he remains out of the fashion. Some years later the public sees through it and takes the sham for what it is worth; it now laughs at it, and the much-admired colour of all these works of fashion falls off like the plaster from a badly-built wall: and they are in the same dilapidated condition. We should be glad and not sorry when a fundamentally wrong notion of which we have been secretly conscious for a long time finally gains a footing and is proclaimed both loudly and openly. The falseness of it will soon be felt and eventually proclaimed equally loudly and openly. It is as if an abscess had burst.

The man who publishes and edits an article written by an anonymous critic should be held as immediately responsible for it as if he had written it himself; just as one holds a manager responsible for bad work done by his workmen. In this way the fellow would be treated as he deserves to be—namely, without any ceremony.

An anonymous writer is a literary fraud against whom one should immediately cry out, "Wretch, if you do not wish to admit what it is you say against other people, hold your slanderous tongue."

An anonymous criticism carries no more weight than an anonymous letter, and should therefore be looked upon with equal mistrust. Or do we wish to accept the assumed name of a man, who in reality represents a *societe anonyme*, as a guarantee for the veracity of his friends?

The little honesty that exists among authors is discernible in the unconscionable way they misquote from the writings of others. I find whole passages in my works wrongly quoted, and it is only in my appendix, which is absolutely lucid, that an exception is made. The misquotation is frequently due to carelessness, the pen of such people has been used to write down such trivial and banal phrases that it goes on writing them out of force of habit. Sometimes the misquotation is due to impertinence on the part of some one who wants to improve upon my work; but a bad motive only too often prompts the misquotation—it is then horrid baseness and roguery, and, like a man who commits forgery, he loses the character for being an honest man for ever.

Style is the physiognomy of the mind. It is a more reliable key to character than the physiognomy of the body. To imitate another person's style is like wearing a mask. However fine the mask, it soon becomes insipid and intolerable because it is without life; so that even the ugliest living face is better. Therefore authors who write in Latin and imitate the style of the old writers essentially wear a mask; one certainly hears what they say, but one cannot watch their physiognomy—that is to say their style. One observes, however, the style in the Latin writings of men *who think for themselves*, those who have not deigned to imitate, as, for instance, Scotus Erigena, Petrarch, Bacon, Descartes, Spinoza, etc.

Affectation in style is like making grimaces. The language in which a man writes is the physiognomy of his nation; it establishes a great many differences, beginning from the language of the Greeks down to that of the Caribbean islanders.

We should seek for the faults in the style of another author's works, so that we may avoid committing the same in our own.

In order to get a provisional estimate of the value of an author's productions it is not exactly necessary to know the matter on which he has thought or what it is he has thought about it,—this would compel one to read the whole of his works,—but it will be sufficient to know *how* he has thought. His *style* is an exact expression of *how* he has thought, of the essential state and general *quality* of his thoughts. It shows the *formal* nature—which must always remain the same—of all the thoughts of a man, whatever the subject on which he has thought or what it is he has said about it. It is the dough out of which all his ideas are kneaded, however various they may be. When Eulenspiegel was asked by a man how long he would have to walk before reaching the next place, and gave the apparently absurd answer *Walk*, his intention was to judge from the man's walking how far he would go in a given time. And so it is when I have read a few pages of an author, I know about how far he can help me.

In the secret consciousness that this is the condition of things, every mediocre writer tries to mask his own natural style. This instantly necessitates his giving up all idea of being *naive*, a privilege which belongs to superior minds sensible of their superiority, and therefore sure of themselves. For instance, it is absolutely impossible for men of ordinary intelligence to make up their minds to write as they think; they resent the idea of their work looking too simple. It would always be of some value, however. If they would only go honestly to work and in a simple way express the few and ordinary ideas they have really thought, they would be readable and even instructive in their own sphere. But instead of that they try to appear to have thought much more deeply than is the case. The result is, they put what they have to say into forced and involved language, create new words and prolix periods which go round the thought and cover it up. They hesitate between the two attempts of communicating the thought and of concealing it. They want to make it look grand so that it has the appearance of being learned and profound, thereby giving one the idea that there is much more in it than one perceives at the moment. Accordingly, they sometimes put down their thoughts in bits, in short, equivocal, and paradoxical sentences which appear to mean much more than they say (a splendid example of this kind of writing is furnished by Schelling's treatises on Natural Philosophy); sometimes they express their thoughts in a crowd of words and the most intolerable diffuseness, as if it were necessary to make a sensation in order to make the profound meaning of their phrases intelligible—while it is quite a simple idea if not a trivial one (examples without number are supplied in Fichte's popular works and in the philosophical pamphlets of a hundred other miserable blockheads that are not worth mentioning), or else they endeavour to use a certain style in writing which it has pleased them to adopt—for example, a style that is so thoroughly *Kat' e'xochae'u* profound and scientific, where one is tortured to death by the narcotic effect of long-spun periods that are void of all thought (examples of this are specially supplied by those most impertinent of all mortals, the Hegelians in their Hegel newspaper commonly known as *Jahrbuecher der wissenschaftlichen Literatur*); or again, they aim at an intellectual style where it seems then as if they wish to go crazy, and so on. All such efforts whereby they try to postpone the *nascetur ridiculus mus* make it frequently difficult to understand what they really mean. Moreover, they write down words, nay, whole periods, which mean nothing in themselves, in the hope, however, that some one else will understand something from them. Nothing else is at the bottom of all such endeavours but the inexhaustible attempt which is always venturing on new paths, to sell words for thoughts, and by means of new expressions, or expressions used in a new sense, turns of phrases and combinations of all kinds, to produce the appearance of intellect in order to compensate for the want of it which is so painfully felt. It is amusing to see how, with this aim in view, first this mannerism and then that is tried; these they intend to represent the mask of intellect: this mask may possibly deceive the inexperienced for a while, until it is recognised as being nothing but a dead mask, when it is laughed at and exchanged for another.

We find a writer of this kind sometimes writing in a dithyrambic style, as if he were intoxicated; at other times, nay, on the very next page, he will be high-sounding, severe, and deeply learned, prolix to the last degree of dulness, and cutting everything very small, like the late Christian Wolf, only in a modern garment. The mask of unintelligibility holds out the longest; this is only in Germany, however, where it was introduced by Fichte, perfected by Schelling, and attained its highest climax finally in Hegel, always with the happiest results. And yet nothing is easier than to write so that no one can understand; on the other hand, nothing is more difficult than to express learned ideas so that every one must understand them. All the arts I have cited above are superfluous if the writer really possesses any intellect, for it allows a man to show himself as he is and verifies for all time what Horace said: *Scribendi recte sapere est et principium et fons.*

But this class of authors is like certain workers in metal, who try a hundred different compositions to take the place of gold, which is the only metal that can never have a substitute. On the contrary, there is nothing an author should guard against more than the apparent endeavour to show more intellect than he has; because this rouses the suspicion in the reader that he has very little, since a man always affects something, be its nature what it may, that he does not really possess. And this is why it is praise to an author to call him naive, for it signifies that he may show himself as he is. In general, naivete attracts, while anything that is unnatural everywhere repels. We also find that every true thinker endeavours to express his thoughts as purely, clearly, definitely, and concisely as ever possible. This is why simplicity has always been looked upon as a token, not only of truth, but also of genius. Style receives its beauty from the thought expressed, while with those writers who only pretend to think it is their thoughts that are said to be fine because of their style. Style is merely the silhouette of thought; and to write in a vague or bad style means a stupid or confused mind.

Hence, the first rule—nay, this in itself is almost sufficient for a good style—is this, *that the author should have something to say*. Ah! this implies a great deal. The neglect of this rule is a fundamental characteristic of the philosophical, and generally speaking of all the reflective authors in Germany, especially since the time of Fichte. It is obvious that all these writers wish *to appear* to have something to say, while they have nothing to say. This mannerism was introduced by the pseudo-philosophers of the Universities and may be discerned everywhere, even among the first literary notabilities of the age. It is the mother of that forced and vague style which seems to have two, nay, many meanings, as well as of that prolix and ponderous style, *le stile empese*; and of that no less useless bombastic style, and finally of that mode of concealing the most awful poverty of thought under a babble of inexhaustible chatter that resembles a clacking mill and is just as stupefying: one may read for hours together without getting hold of a single clearly defined and definite idea. The *Halleschen*, afterwards called the *Deutschen Jahrbuecher*, furnishes almost throughout excellent examples of this style of writing. The Germans, by the way, from force of habit read page after page of all kinds of such verbiage without getting any definite idea of what the author really means: they think it all very proper and do not discover that he is writing merely for the sake of writing. On the other hand, a good author who is rich in ideas soon gains the reader's credit of having really and truly *something to say*; and this gives the intelligent reader patience to follow him attentively. An author of this kind will always express himself in the simplest and most direct manner, for the very reason that he really has something to say; because he wishes to awaken in the reader the same idea he has in his own mind and no other. Accordingly he will be able to say with Boileau—

"Ma pensee au grand jour partout s'offre et s'expose,
Et mon vers, bien ou mal, dit toujours quelque chose;"

while of those previously described writers it may be said, in the words of the same poet, *et qui parlant beaucoup ne disent jamais rien*. It is also a characteristic of such writers to avoid, if it is possible, expressing themselves *definitely*, so that they may be always able in case of need to get out of a difficulty; this is why they always choose the more *abstract* expressions: while people of intellect choose the more *concrete*; because the latter bring the matter closer to view, which is the source of all evidence. This preference for abstract expressions may be confirmed by numerous examples: a specially ridiculous example is the following. Throughout German literature of the last ten years we find "to condition" almost everywhere used in place of "to cause" or "to effect." Since it is more abstract and indefinite it says less than it implies, and consequently leaves a little back door open to please those whose secret consciousness of their own incapacity inspires them with a continual fear of all *definite* expressions. While with other people it is merely the effect of that national tendency to immediately imitate everything that is stupid in literature and wicked in life; this is shown in either case by the quick way in which it spreads. The Englishman depends on his own judgment both in what he writes and what he does, but this applies less to the German than to any other nation. In consequence of the state of things referred to, the words "to cause" and "to effect" have almost entirely disappeared from the literature of the last ten years, and people everywhere talk of "to condition." The fact is worth mentioning because it is characteristically ridiculous. Everyday authors are only half conscious when they write, a fact which accounts for their want of intellect and the tediousness of their writings; they do not really themselves understand the meaning of their own words, because they take ready-made words and learn them. Hence they combine whole phrases more than words—*phrases banales*. This accounts for that obviously characteristic want of clearly defined thought; in fact, they lack the die that stamps their thoughts, they have no clear thought of their own; in place of it we find an indefinite, obscure interweaving of words, current phrases, worn-out terms of speech, and

43

fashionable expressions. The result is that their foggy kind of writing is like print that has been done with old type. On the other hand, intelligent people *really* speak to us in their writings, and this is why they are able to both move and entertain us. It is only intelligent writers who place individual words together with a full consciousness of their use and select them with deliberation. Hence their style of writing bears the same relation to that of those authors described above, as a picture that is really painted does to one that has been executed with stencil. In the first instance every word, just as every stroke of the brush, has some special significance, while in the other everything is done mechanically. The same distinction may be observed in music. For it is the omnipresence of intellect that always and everywhere characterises the works of the genius; and analogous to this is Lichtenberg's observation, namely, that Garrick's soul was omnipresent in all the muscles of his body. With regard to the tediousness of the writings referred to above, it is to be observed in general that there are two kinds of tediousness—an objective and a subjective. The *objective* form of tediousness springs from the deficiency of which we have been speaking—that is to say, where the author has no perfectly clear thought or knowledge to communicate. For if a writer possesses any clear thought or knowledge it will be his aim to communicate it, and he will work with this end in view; consequently the ideas he furnishes are everywhere clearly defined, so that he is neither diffuse, unmeaning, nor confused, and consequently not tedious. Even if his fundamental idea is wrong, yet in such a case it will be clearly thought out and well pondered; in other words, it is at least formally correct, and the writing is always of some value. While, for the same reason, a work that is objectively *tedious* is at all times without value. Again, *subjective* tediousness is merely relative: this is because the reader is not interested in the subject of the work, and that what he takes an interest in is of a very limited nature. The most excellent work may therefore be tedious subjectively to this or that person, just as, *vice versa*, the worst work may be subjectively diverting to this or that person: because he is interested in either the subject or the writer of the book.

It would be of general service to German authors if they discerned that while a man should, if possible, think like a great mind, he should speak the same language as every other person. Men should use common words to say uncommon things, but they do the reverse. We find them trying to envelop trivial ideas in grand words and to dress their very ordinary thoughts in the most extraordinary expressions and the most outlandish, artificial, and rarest phrases. Their sentences perpetually stalk about on stilts. With regard to their delight in bombast, and to their writing generally in a grand, puffed-up, unreal, hyperbolical, and acrobatic style, their prototype is Pistol, who was once impatiently requested by Falstaff, his friend, to "say what you have to say, *like a man of this world*."[5]

There is no expression in the German language exactly corresponding to *stile empese*; but the thing itself is all the more prevalent. When combined with unnaturalness it is in works what affected gravity, grandness, and unnaturalness are in social intercourse; and it is just as intolerable. Poverty of intellect is fond of wearing this dress; just as stupid people in everyday life are fond of assuming gravity and formality.

A man who writes in this *prezioes* style is like a person who dresses himself up to avoid being mistaken for or confounded with the mob; a danger which a *gentleman*, even in his worst clothes, does not run. Hence just as a plebeian is recognised by a certain display in his dress and his *tire a quatre epingles*, so is an ordinary writer recognised by his style.

If a man has something to say that is worth saying, he need not envelop it in affected expressions, involved phrases, and enigmatical innuendoes; but he may rest assured that by expressing himself in a simple, clear, and naive manner he will not fail to produce the right effect. A man who makes use of such artifices as have been alluded to betrays his poverty of ideas, mind, and knowledge.

Nevertheless, it is a mistake to attempt to write exactly as one speaks. Every style of writing should bear a certain trace of relationship with the

45

monumental style, which is, indeed, the ancestor of all styles; so that to write as one speaks is just as faulty as to do the reverse, that is to say, to try and speak as one writes. This makes the author pedantic, and at the same time difficult to understand.

Obscurity and vagueness of expression are at all times and everywhere a very bad sign. In ninety-nine cases out of a hundred they arise from vagueness of thought, which, in its turn, is almost always fundamentally discordant, inconsistent, and therefore wrong. When a right thought springs up in the mind it strives after clearness of expression, and it soon attains it, for clear thought easily finds its appropriate expression. A man who is capable of thinking can express himself at all times in clear, comprehensible, and unambiguous words. Those writers who construct difficult, obscure, involved, and ambiguous phrases most certainly do not rightly know what it is they wish to say: they have only a dull consciousness of it, which is still struggling to put itself into thought; they also often wish to conceal from themselves and other people that in reality they have nothing to say. Like Fichte, Schelling, and Hegel, they wish to appear to know what they do not know, to think what they do not think, and to say what they do not say.

Will a man, then, who has something real to impart endeavour to say it in a clear or an indistinct way? Quintilian has already said, *plerumque accidit ut faciliora sint ad intelligendum et lucidiora multo, quae a doctissimo quoque dicuntur . . . Erit ergo etiam obscurior, quo quisque deterior.*

A man's way of expressing himself should not be *enigmatical*, but he should know whether he has something to say or whether he has not. It is an uncertainty of expression which makes German writers so dull. The only exceptional cases are those where a man wishes to express something that is in some respect of an illicit nature. As anything that is far-fetched generally produces the reverse of what the writer has aimed at, so do words serve to make thought comprehensible; but only up to a certain point. If words are piled up beyond this point they make the thought that is being communicated more and more obscure. To hit that point is the problem of style and a matter of discernment; for every superfluous word prevents its purpose being carried out. Voltaire means this when he says: *l'adjectif est l'ennemi du substantif.* (But, truly, many authors try to hide their poverty of thought under a superfluity of words.)

Accordingly, all prolixity and all binding together of unmeaning observations that are not worth reading should be avoided. A writer must be sparing with the reader's time, concentration, and patience; in this way he makes him believe that what he has before him is worth his careful reading, and will repay the trouble he has spent upon it. It is always better to leave out something that is good than to write down something that is not worth saying. Hesiod's [Greek: pleon haemisu pantos][6] finds its right application. In fact, not to say everything! *Le secret pour etre ennuyeux, c'est de tout dire.* Therefore, if possible, the quintessence only! the chief matter only! nothing that the reader would think for himself. The use of many words in order to express little thought is everywhere the infallible sign of mediocrity; while to clothe much thought in a few words is the infallible sign of distinguished minds.

Truth that is naked is the most beautiful, and the simpler its expression the deeper is the impression it makes; this is partly because it gets unobstructed hold of the hearer's mind without his being distracted by secondary thoughts, and partly because he feels that here he is not being corrupted or deceived by the arts of rhetoric, but that the whole effect is got from the thing itself. For instance, what declamation on the emptiness of human existence could be more impressive than Job's: *Homo, natus de muliere, brevi vivit tempore, repletus multis miseriis, qui, tanquam flos, egreditur et conteritur, et fugit velut umbra.* It is for this very reason that the naive poetry of Goethe is so incomparably greater than the rhetorical of Schiller. This is also why many folk-songs have so great an effect upon us. An author should guard against using all unnecessary rhetorical adornment, all useless amplification, and in general, just as in architecture he should guard against an excess of decoration, all superfluity of expression—in other words, he must aim at *chastity* of style. Everything that is redundant has a harmful effect. The law of simplicity and naivete applies to all fine art, for it is compatible with what is most sublime.

True brevity of expression consists in a man only saying what is worth saying, while avoiding all diffuse explanations of things which every one can think out for himself; that is, it consists in his correctly distinguishing between what is necessary and what is superfluous. On the other hand, one should never sacrifice clearness, to say nothing of grammar, for the sake of being brief. To impoverish the expression of a thought, or to obscure or spoil the meaning of a period for the sake of using fewer words shows a lamentable want of judgment. And this is precisely what that false brevity nowadays in vogue is trying to do, for writers not only leave out words that are to the purpose, but even grammatical and logical essentials.[7]

Subjectivity, which is an error of style in German literature, is, through the deteriorated condition of literature and neglect of old languages, becoming more common. By *subjectivity* I mean when a writer thinks it sufficient for himself to know what he means and wants to say, and it is left to the reader to discover what is meant. Without troubling himself about his reader, he writes as if he were holding a monologue; whereas it should be a dialogue, and, moreover, a dialogue in which he must express himself all the more clearly as the questions of the reader cannot be heard. And it is for this very reason that style should not be subjective but objective, and for it to be objective the words must be written in such a way as to directly compel the reader to think precisely the same as the author thought. This will only be the case when the author has borne in mind that thoughts, inasmuch as they follow the law of gravity, pass more easily from head to paper than from paper to head. Therefore the journey from paper to head must be helped by every means at his command. When he does this his words have a purely objective effect, like that of a completed oil painting; while the subjective style is not much more certain in its effect than spots on the wall, and it is only the man whose fantasy is accidentally aroused by them that sees figures; other people only see blurs. The difference referred to applies to every style of writing as a whole, and it is also often met with in particular instances; for example, I read in a book that has just been published: *I have not written to increase the number of existing books.* This means exactly the opposite of what the writer had in view, and is nonsense into the bargain.

A man who writes carelessly at once proves that he himself puts no great value on his own thoughts. For it is only by being convinced of the truth and importance of our thoughts that there arises in us the inspiration necessary for the inexhaustible patience to discover the clearest, finest, and most powerful expression for them; just as one puts holy relics or priceless works of art in silvern or golden receptacles. It was for this reason that the old writers—whose thoughts, expressed in their own words, have lasted for thousands of years and hence bear the honoured title of classics—wrote with universal care. Plato, indeed, is said to have written the introduction to his *Republic* seven times with different modifications. On the other hand, the Germans are conspicuous above all other nations for neglect of style in writing, as they are for neglect of dress, both kinds of slovenliness which have their source in the German national character. Just as neglect of dress betrays contempt for the society in which a man moves, so does a hasty, careless, and bad style show shocking disrespect for the reader, who then rightly punishes it by not reading the book.

FOOTNOTES:

5   Schopenhauer here gives an example of this bombastic style which would be of little interest to English readers.—TRANSLATOR.

6   *Opera et dies*, v. 40.

7   Schopenhauer here at length points out various common errors in the writing and speaking of German which would lose significance in a translation.—TR.

ON NOISE.

Kant has written a treatise on *The Vital Powers*; but I should like to write a dirge on them, since their lavish use in the form of knocking, hammering, and tumbling things about has made the whole of my life a daily torment. Certainly there are people, nay, very many, who will smile at this, because they are not sensitive to noise; it is precisely these people, however, who are not sensitive to argument, thought, poetry or art, in short, to any kind of intellectual impression: a fact to be assigned to the coarse quality and strong texture of their brain tissues. On the other hand, in the biographies or in other records of the personal utterances of almost all great writers, I find complaints of the pain that noise has occasioned to intellectual men. For example, in the case of Kant, Goethe, Lichtenberg, Jean Paul; and indeed when no mention is made of the matter it is merely because the context did not lead up to it. I should explain the subject we are treating in this way: If a big diamond is cut up into pieces, it immediately loses its value as a whole; or if an army is scattered or divided into small bodies, it loses all its power; and in the same way a great intellect has no more power than an ordinary one as soon as it is interrupted, disturbed, distracted, or diverted; for its superiority entails that it concentrates all its strength on one point and object, just as a concave mirror concentrates all the rays of light thrown upon it. Noisy interruption prevents this concentration. This is why the most eminent intellects have always been strongly averse to any kind of disturbance, interruption and distraction, and above everything to that violent interruption which is caused by noise; other people do not take any particular notice of this sort of thing. The most intelligent of all the European nations has called "Never interrupt" the eleventh commandment. But noise is the most impertinent of all interruptions, for it not only interrupts our own thoughts but disperses them. Where, however, there is nothing to interrupt, noise naturally will not be felt particularly. Sometimes a trifling but incessant noise torments and disturbs me for a time, and before I become distinctly conscious of it I feel it merely as the effort of thinking becomes more difficult, just as I should feel a weight on my foot; then I realise what it is.

But to pass from *genus* to *species*, the truly infernal cracking of whips in the narrow resounding streets of a town must be denounced as the most unwarrantable and disgraceful of all noises. It deprives life of all peace and sensibility. Nothing gives me so clear a grasp of the stupidity and thoughtlessness of mankind as the tolerance of the cracking of whips. This sudden, sharp crack which paralyses the brain, destroys all meditation, and murders thought, must cause pain to any one who has anything like an idea in his head. Hence every crack must disturb a hundred people applying their minds to some activity, however trivial it may be; while it disjoints and renders painful the meditations of the thinker; just like the executioner's axe when it severs the head from the body. No sound cuts so sharply into the brain as this cursed cracking of whips; one feels the prick of the whip-cord in one's brain, which is affected in the same way as the *mimosa pudica* is by touch, and which lasts the same length of time. With all respect for the most holy doctrine of utility, I do not see why a fellow who is removing a load of sand or manure should obtain the privilege of killing in the bud the thoughts that are springing up in the heads of about ten thousand people successively. (He is only half-an-hour on the road.)

Hammering, the barking of dogs, and the screaming of children are abominable; but it is *only* the cracking of a whip that is the true murderer of thought. Its object is to destroy every favourable moment that one now and then may have for reflection. If there were no other means of urging on an animal than by making this most disgraceful of all noises, one would forgive its existence. But it is quite the contrary: this cursed cracking of whips is not only unnecessary but even useless. The effect that it is intended to have on the horse mentally becomes quite blunted and ineffective; since the constant abuse of it has accustomed the horse to the crack, he does not quicken his pace for it. This is especially noticeable in the unceasing crack of the whip which comes from an empty vehicle as it is being driven at its slowest rate to pick up a fare. The slightest touch with the whip would be more effective. Allowing, however, that it were absolutely necessary to remind the horse of the presence of the whip by continually cracking it, a crack that made one hundredth part of the noise would be sufficient. It is well known that animals in regard to hearing and seeing notice the slightest indications, even indications that are scarcely perceptible to ourselves. Trained dogs and canary birds furnish astonishing examples of this. Accordingly, this cracking of whips must be regarded as something purely wanton; nay, as an impudent defiance, on the part of those who work with their hands, offered to those who work with their heads. That such infamy is endured in a town is a piece of barbarity and injustice, the more so as it could be easily removed by a police notice requiring every whip cord to have a knot at the end of it. It would do no harm to draw the proletariat's attention to the classes above him who work with their heads; for he has unbounded fear of any kind of head work. A fellow who rides through the narrow streets of a populous town with unemployed post-horses or cart-horses, unceasingly cracking with all his strength a whip several yards long, instantly deserves to dismount and receive five really good blows with a stick. If all the philanthropists in the world, together with all the legislators, met in order to bring forward their reasons for the total abolition of corporal punishment, I would not be persuaded to the contrary.

But we can see often enough something that is even still worse. I mean a carter walking alone, and without any horses, through the streets incessantly cracking his whip. He has become so accustomed to the crack in consequence of its unwarrantable toleration. Since one looks after one's body and all its needs in a most tender fashion, is the thinking mind to be the only thing that never experiences the slightest consideration or protection, to say nothing of respect? Carters, sack-bearers (porters), messengers, and such-like, are the beasts of burden of humanity; they should be treated absolutely with justice, fairness, forbearance and care, but they ought not to be allowed to thwart the higher exertions of the human race by wantonly making a noise. I should like to know how many great and splendid thoughts these whips have cracked out of the world. If I had any authority, I should soon produce in the heads of these carters an inseparable *nexus idearum* between cracking a whip and receiving a whipping.

Let us hope that those nations with more intelligence and refined feelings will make a beginning, and then by force of example induce the Germans to do the same.[8] Meanwhile, hear what Thomas Hood says of them *(Up the Rhine)*: *"For a musical people they are the most noisy I ever met with."* That they are so is not due to their being more prone to making a noise than other people, but to their insensibility, which springs from obtuseness; they are not disturbed by it in reading or thinking, because they do not think; they only smoke, which is their substitute for thought. The general toleration of unnecessary noise, for instance, of the clashing of doors, which is so extremely ill-mannered and vulgar, is a direct proof of the dulness and poverty of thought that one meets with everywhere. In Germany it seems as though it were planned that no one should think for noise; take the inane drumming that goes on as an instance. Finally, as far as the literature treated of in this chapter is concerned, I have only one work to recommend, but it is an excellent one: I mean a poetical epistle in *terzo rimo* by the famous painter Bronzino, entitled *"De' Romori: a Messer Luca Martini."* It describes fully and amusingly the torture to which one is put by the many kinds of noises of a small Italian town. It is written in tragicomic style. This epistle is to be found in *Opere burlesche del Berni, Aretino ed altri,* vol. ii. p. 258, apparently published in Utrecht in 1771.

The nature of our intellect is such that *ideas* are said to spring by abstraction from *observations*, so that the latter are in existence before the former. If this is really what takes place, as is the case with a man who has merely his own experience as his teacher and book, he knows quite well which of his observations belong to and are represented by each of his ideas; he is perfectly acquainted with both, and accordingly he treats everything correctly that comes before his notice. We might call this the natural mode of education.

On the other hand, an artificial education is having one's head crammed full of ideas, derived from hearing others talk, from learning and reading, before one has anything like an extensive knowledge of the world as it is and as one sees it. The observations which produce all these ideas are said to come later on with experience; but until then these ideas are applied wrongly, and accordingly both things and men are judged wrongly, seen wrongly, and treated wrongly. And so it is that education perverts the mind; and this is why, after a long spell of learning and reading, we enter the world, in our youth, with views that are partly simple, partly perverted; consequently we comport ourselves with an air of anxiety at one time, at another of presumption. This is because our head is full of ideas which we are now trying to make use of, but almost always apply wrongly. This is the result of [Greek: hysteron proteron] (putting the cart before the horse), since we are directly opposing the natural development of our mind by obtaining ideas first and observations last; for teachers, instead of developing in a boy his faculties of discernment and judgment, and of thinking for himself, merely strive to stuff his head full of other people's thoughts. Subsequently, all the opinions that have sprung from misapplied ideas have to be rectified by a lengthy experience; and it is seldom that they are completely rectified. This is why so few men of learning have such sound common sense as is quite common among the illiterate.

* * * * *

From what has been said, the principal point in education is that *one's knowledge of the world begins at the right end;* and the attainment of which might be designated as the aim of all education. But, as has been pointed out, this depends principally on the observation of each thing preceding the idea one forms of it; further, that narrow ideas precede broader; so that the whole of one's instruction is given in the order that the ideas themselves during formation must have followed. But directly this order is not strictly adhered to, imperfect and subsequently wrong ideas spring up; and finally there arises a perverted view of the world in keeping with the nature of the individual—a view such as almost every one holds for a long time, and most people to the end of their lives. If a man analyses his own character, he will find that it was not until he reached a very ripe age, and in some cases quite unexpectedly, that he was able to rightly and clearly understand many matters of a quite simple nature.

Previously, there had been an obscure point in his knowledge of the world which had arisen through his omitting something in his early education, whether he had been either artificially educated by men or just naturally by his own experience. Therefore one should try to find out the strictly natural course of knowledge, so that by keeping methodically to it children may become acquainted with the affairs of the world, without getting false ideas into their heads, which frequently cannot be driven out again. In carrying this out, one must next take care that children do not use words with which they connect no clear meaning. Even children have, as a rule, that unhappy tendency of being satisfied with words instead of wishing to understand things, and of learning words by heart, so that they may make use of them when they are in a difficulty. This tendency clings to them afterwards, so that the knowledge of many learned men becomes mere verbosity.

However, the principal thing must always be to let one's observations precede one's ideas, and not the reverse as is usually and unfortunately the case; which may be likened to a child coming into the world with its feet foremost, or a rhyme begun before thinking of its reason. While the child's mind has made a very few observations one inculcates it with ideas and opinions, which are, strictly speaking, prejudices. His observations and experience are developed through this ready-made apparatus instead of his ideas being developed out of his own observations. In viewing the world one sees many things from many sides, consequently this is not such a short or quick way of learning as that which makes use of abstract ideas, and quickly comes to a decision about everything; therefore preconceived ideas will not be rectified until late, or it may be they are never rectified. For, when a man's view contradicts his ideas, he will reject at the outset what it renders evident as one-sided, nay, he will deny it and shut his eyes to it, so that his preconceived ideas may remain unaffected. And so it happens that many men go through life full of oddities, caprices, fancies, and prejudices, until they finally become fixed ideas. He has never attempted to abstract fundamental ideas from his own observations and experience, because he has got everything ready-made from other people; and it is for this very reason that he and countless others are so insipid and shallow. Instead of such a system, the natural system of education should be employed in educating children. No idea should be impregnated but what has come through the medium of observations, or at any rate been verified by them. A child would have fewer ideas, but they would be well-grounded and correct. It would learn to measure things according to its own standard and not according to another's. It would then never acquire a thousand whims and prejudices which must be eradicated by the greater part of subsequent experience and education. Its mind would henceforth be accustomed to thoroughness and clearness; the child would rely on its own judgment, and be free from prejudices. And, in general, children should not get to know life, in any aspect whatever, from the copy before they have learnt it from the original. Instead, therefore, of hastening to place mere books in their hands, one should make them gradually acquainted with things and the circumstances of human life, and above everything one should take care to guide them to a clear grasp of reality, and to teach them to obtain their ideas directly from the real world, and to form them in keeping with it—but not to get them from elsewhere, as from books, fables, or what others have said—and then later to make use of such ready-made ideas in real life. The result will be that their heads are full of chimeras and that some will have a wrong comprehension of things, and others will fruitlessly endeavour to remodel the world according to those chimeras, and so get on to wrong paths both in theory and practice. For it is incredible how much harm is done by false notions which have been implanted early in life, only to develop later on into prejudices; the later education which we get from the world and real life must be employed in eradicating these early ideas. And this is why, as is related by Diogenes Laertius, Antisthenes gave the following answer: [Greek: erotaetheis ti ton mathaematon anankaiotaton, ephae, "to kaka apomathein."] *(Interrogatus quaenam esset disciplina maxime necessaria, Mala, inquit, dediscere.)*

* * * * *

Children should be kept from all kinds of instruction that may make errors possible until their sixteenth year, that is to say, from philosophy, religion, and general views of every description; because it is the errors that are acquired in early days that remain, as a rule, ineradicable, and because the faculty of judgment is the last to arrive at maturity. They should only be interested in such things that make errors impossible, such as mathematics, in things which are not very dangerous, such as languages, natural science, history, and so forth; in general, the branches of knowledge which are to be taken up at any age must be within reach of the intellect at that age and perfectly comprehensible to it. Childhood and youth are the time for collecting data and getting to know specially and thoroughly individual and particular things. On the other hand, all judgment of a general nature must at that time be suspended, and final explanations left alone. One should leave the faculty of judgment alone, as it only comes with maturity and experience, and also take care that one does not anticipate it by inculcating prejudice, when it will be crippled for ever.

On the contrary, the memory is to be specially exercised, as it has its greatest strength and tenacity in youth; however, what has to be retained must be chosen with the most careful and scrupulous consideration. For as it is what we have learnt well in our youth that lasts, we should take the greatest possible advantage of this precious gift. If we picture to ourselves how deeply engraven on our memory the people are whom we knew during the first twelve years of our life, and how indelibly imprinted are also the events of that time, and most of the things that we then experienced, heard, or learnt, the idea of basing education on this susceptibility and tenacity of the youthful mind will seem natural; in that the mind receives its impressions according to a strict method and a regular system. But because the years of youth that are assigned to man are only few, and the capacity for remembering, in general, is always limited (and still more so the capacity for remembering of the individual), everything depends on the memory being filled with what is most essential and important in any department of knowledge, to the exclusion of everything else. This selection should be made by the most capable minds and masters in every branch of knowledge after the most mature consideration, and the result of it established. Such a selection must be based on a sifting of matters which are necessary and important for a man to know in general, and also for him to know in a particular profession or calling. Knowledge of the first kind would have to be divided into graduated courses, like an encyclopaedia, corresponding to the degree of general culture which each man has attained in his external circumstances; from a course restricted to what is necessary for primary instruction up to the matter contained in every branch of the philosophical faculty. Knowledge of the second kind would, however, be reserved for him who had really mastered the selection in all its branches. The whole would give a canon specially devised for intellectual education, which naturally would require revision every ten years. By such an arrangement the youthful power of

58

the memory would be put to the best advantage, and it would furnish the faculty of judgment with excellent material when it appeared later on.

* * * * *

What is meant by maturity of knowledge is that state of perfection to which any one individual is able to bring it, when an exact correspondence has been effected between the whole of his abstract ideas and his own personal observations: whereby each of his ideas rests directly or indirectly on a basis of observation, which alone gives it any real value; and likewise he is able to place every observation that he makes under the right idea corresponding to it.

Maturity of knowledge is the work of experience alone, and consequently of time. For the knowledge we acquire from our own observation is, as a rule, distinct from that we get through abstract ideas; the former is acquired in the natural way, while the latter comes through good and bad instruction and what other people have told to us. Consequently, in youth there is generally little harmony and connection between our ideas, which mere expressions have fixed, and our real knowledge, which has been acquired by observation. Later they both gradually approach and correct each other; but maturity of knowledge does not exist until they have become quite incorporated. This maturity is quite independent of that other kind of perfection, the standard of which may be high or low, I mean the perfection to which the capacities of an individual may be brought; it is not based on a correspondence between the abstract and intuitive knowledge, but on the degree of intensity of each.

The most necessary thing for the practical man is the attainment of an exact and thorough knowledge of *what is really going on in the world;* but it is also the most irksome, for a man may continue studying until old age without having learnt all that is to be learnt; while one can master the most important things in the sciences in one's youth. In getting such a knowledge of the world, it is as a novice that the boy and youth have the first and most difficult lessons to learn; but frequently even the matured man has still much to learn. The study is of considerable difficulty in itself, but it is made doubly difficult by *novels*, which depict the ways of the world and of men who do not exist in real life. But these are accepted with the credulity of youth, and become incorporated with the mind; so that now, in the place of purely negative ignorance, a whole framework of wrong ideas, which are positively wrong, crops up, subsequently confusing the schooling of experience and representing the lesson it teaches in a false light. If the youth was previously in the dark, he will now be led astray by a will-o'-the-wisp: and with a girl this is still more frequently the case. They have been deluded into an absolutely false view of life by reading novels, and expectations have been raised that can never be fulfilled. This generally has the most harmful effect on their whole lives. Those men who had neither time nor opportunity to read novels in their youth, such as those who work with their hands, have decided advantage over them. Few of these novels are exempt from reproach—nay, whose effect is contrary to bad. Before all others, for instance, *Gil Blas* and the other works of Le Sage (or rather their Spanish originals); further, *The Vicar of Wakefield*, and to some extent the novels of Walter Scott. *Don Quixote* may be regarded as a satirical presentation of the error in question.

FOOTNOTES:

8   According to a notice from the Munich Society for the Protection of Animals, the superfluous whipping and cracking were strictly forbidden in Nuremberg in December 1858.

ON READING AND BOOKS.

Ignorance is degrading only when it is found in company with riches. Want and penury restrain the poor man; his employment takes the place of knowledge and occupies his thoughts: while rich men who are ignorant live for their pleasure only, and resemble a beast; as may be seen daily. They are to be reproached also for not having used wealth and leisure for that which lends them their greatest value.

When we read, another person thinks for us: we merely repeat his mental process. It is the same as the pupil, in learning to write, following with his pen the lines that have been pencilled by the teacher. Accordingly, in reading, the work of thinking is, for the greater part, done for us. This is why we are consciously relieved when we turn to reading after being occupied with our own thoughts. But, in reading, our head is, however, really only the arena of some one else's thoughts. And so it happens that the person who reads a great deal—that is to say, almost the whole day, and recreates himself by spending the intervals in thoughtless diversion, gradually loses the ability to think for himself; just as a man who is always riding at last forgets how to walk. Such, however, is the case with many men of learning: they have read themselves stupid. For to read in every spare moment, and to read constantly, is more paralysing to the mind than constant manual work, which, at any rate, allows one to follow one's own thoughts. Just as a spring, through the continual pressure of a foreign body, at last loses its elasticity, so does the mind if it has another person's thoughts continually forced upon it. And just as one spoils the stomach by overfeeding and thereby impairs the whole body, so can one overload and choke the mind by giving it too much nourishment. For the more one reads the fewer are the traces left of what one has read; the mind is like a tablet that has been written over and over. Hence it is impossible to reflect; and it is only by reflection that one can assimilate what one has read if one reads straight ahead without pondering over it later, what has been read does not take root, but is for the most part lost. Indeed, it is the same with mental as with bodily food: scarcely the fifth part of what a man takes is assimilated; the remainder passes off in evaporation, respiration, and the like.

From all this it may be concluded that thoughts put down on paper are nothing more than footprints in the sand: one sees the road the man has taken, but in order to know what he saw on the way, one requires his eyes.

*   *   *   *   *

No literary quality can be attained by reading writers who possess it: be it, for example, persuasiveness, imagination, the gift of drawing comparisons, boldness or bitterness, brevity or grace, facility of expression or wit, unexpected contrasts, a laconic manner, naivete, and the like. But if we are already gifted with these qualities—that is to say, if we possess them *potentia*—we can call them forth and bring them to consciousness; we can discern to what uses they are to be put; we can be strengthened in our inclination, nay, may have courage, to use them; we can judge by examples the effect of their application and so learn the correct use of them; and it is only after we have accomplished all this that we *actu* possess these qualities. This is the only way in which reading can form writing, since it teaches us the use to which we can put our own natural gifts; and in order to do this it must be taken for granted that these qualities are in us. Without them we learn nothing from reading but cold, dead mannerisms, and we become mere imitators.

*   *   *   *   *

The health officer should, in the interest of one's eyes, see that the smallness of print has a fixed minimum, which must not be exceeded. When I was in Venice in 1818, at which time the genuine Venetian chain was still being made, a goldsmith told me that those who made the *catena fina* turned blind at thirty.

* * * * *

As the strata of the earth preserve in rows the beings which lived in former times, so do the shelves of a library preserve in a like manner the errors of the past and expositions concerning them. Like those creatures, they too were full of life in their time and made a great deal of noise; but now they are stiff and fossilised, and only of interest to the literary palaeontologist.

* * * * *

According to Herodotus, Xerxes wept at the sight of his army, which was too extensive for him to scan, at the thought that a hundred years hence not one of all these would be alive. Who would not weep at the thought in looking over a big catalogue that of all these books not one will be in existence in ten years' time?

It is the same in literature as in life. Wherever one goes one immediately comes upon the incorrigible mob of humanity. It exists everywhere in legions; crowding, soiling everything, like flies in summer. Hence the numberless bad books, those rank weeds of literature which extract nourishment from the corn and choke it.

They monopolise the time, money, and attention which really belong to good books and their noble aims; they are written merely with a view to making money or procuring places. They are not only useless, but they do positive harm. Nine-tenths of the whole of our present literature aims solely at taking a few shillings out of the public's pocket, and to accomplish this, author, publisher, and reviewer have joined forces.

There is a more cunning and worse trick, albeit a profitable one. *Litterateurs*, hack-writers, and productive authors have succeeded, contrary to good taste and the true culture of the age, in bringing the world *elegante* into leading-strings, so that they have been taught to read *a tempo* and all the same thing—namely, *the newest books* order that they may have material for conversation in their social circles. Bad novels and similar productions from the pen of writers who were once famous, such as Spindler, Bulwer, Eugene Sue, and so on, serve this purpose. But what can be more miserable than the fate of a reading public of this kind, that feels always impelled to read the latest writings of extremely commonplace authors who write for money only, and therefore exist in numbers? And for the sake of this they merely know by name the works of the rare and superior writers, of all ages and countries.

Literary newspapers, since they print the daily smatterings of commonplace people, are especially a cunning means for robbing from the aesthetic public the time which should be devoted to the genuine productions of art for the furtherance of culture.

Hence, in regard to our subject, the art of *not* reading is highly important. This consists in not taking a book into one's hand merely because it is interesting the great public at the time—such as political or religious pamphlets, novels, poetry, and the like, which make a noise and reach perhaps several editions in their first and last years of existence. Remember rather that the man who writes for fools always finds a large public: and only read for a limited and definite time exclusively the works of great minds, those who surpass other men of all times and countries, and whom the voice of fame points to as such. These alone really educate and instruct.

One can never read too little of bad, or too much of good books: bad books are intellectual poison; they destroy the mind.

In order to read what is good one must make it a condition never to read what is bad; for life is short, and both time and strength limited.

* * * * *

Books are written sometimes about this, sometimes about that great thinker of former times, and the public reads these books, but not the works of the man himself. This is because it wants to read only what has just been printed, and because *similis simili gaudet*, and it finds the shallow, insipid gossip of some stupid head of to-day more homogeneous and agreeable than the thoughts of great minds. I have to thank fate, however, that a fine epigram of A.B. Schlegel, which has since been my guiding star, came before my notice as a youth:

"Leset fleizig die Alten, die wahren eigentlich Alten
Was die Neuen davon sagen bedeutet nicht viel."

Oh, how like one commonplace mind is to another! How they are all fashioned in one form! How they all think alike under similar circumstances, and never differ! This is why their views are so personal and petty. And a stupid public reads the worthless trash written by these fellows for no other reason than that it has been printed to-day, while it leaves the works of great thinkers undisturbed on the bookshelves.

Incredible are the folly and perversity of a public that will leave unread writings of the noblest and rarest of minds, of all times and all countries, for the sake of reading the writings of commonplace persons which appear daily, and breed every year in countless numbers like flies; merely because these writings have been printed to-day and are still wet from the press. It would be better if they were thrown on one side and rejected the day they appeared, as they must be after the lapse of a few years. They will then afford material for laughter as illustrating the follies of a former time.

It is because people will only read what is *the newest* instead of what is the best of all ages, that writers remain in the narrow circle of prevailing ideas, and that the age sinks deeper and deeper in its own mire.

* * * * *

There are at all times two literatures which, although scarcely known to each other, progress side by side—the one real, the other merely apparent. The former grows into literature that *lasts*. Pursued by people who live *for* science or poetry, it goes its way earnestly and quietly, but extremely slowly; and it produces in Europe scarcely a dozen works in a century, which, however, are *permanent*. The other literature is pursued by people who live *on* science or poetry; it goes at a gallop amid a great noise and shouting of those taking part, and brings yearly many thousand works into the market. But after a few years one asks, Where are they? where is their fame, which was so great formerly? This class of literature may be distinguished as fleeting, the other as permanent.

* * * * *

It would be a good thing to buy books if one could also buy the time to read them; but one usually confuses the purchase of books with the acquisition of their contents. To desire that a man should retain everything he has ever read, is the same as wishing him to retain in his stomach all that he has ever eaten. He has been bodily nourished on what he has eaten, and mentally on what he has read, and through them become what he is. As the body assimilates what is homogeneous to it, so will a man *retain* what *interests* him; in other words, what coincides with his system of thought or suits his ends. Every one has aims, but very few have anything approaching a system of thought. This is why such people do not take an objective interest in anything, and why they learn nothing from what they read: they remember nothing about it.

Repetitio est mater studiorum. Any kind of important book should immediately be read twice, partly because one grasps the matter in its entirety the second time, and only really understands the beginning when the end is known; and partly because in reading it the second time one's temper and mood are different, so that one gets another impression; it may be that one sees the matter in another light.

Works are the quintessence of a mind, and are therefore always of by far greater value than conversation, even if it be the conversation of the greatest mind. In every essential a man's works surpass his conversation and leave it far behind. Even the writings of an ordinary man may be instructive, worth reading, and entertaining, for the simple reason that they are the quintessence of that man's mind—that is to say, the writings are the result and fruit of his whole thought and study; while we should be dissatisfied with his conversation. Accordingly, it is possible to read books written by people whose conversation would give us no satisfaction; so that the mind will only by degrees attain high culture by finding entertainment almost entirely in books, and not in men.

There is nothing that so greatly recreates the mind as the works of the old classic writers. Directly one has been taken up, even if it is only for half-an-hour, one feels as quickly refreshed, relieved, purified, elevated, and strengthened as if one had refreshed oneself at a mountain stream. Is this due to the perfections of the old languages, or to the greatness of the minds whose works have remained unharmed and untouched for centuries? Perhaps to both combined. This I know, directly we stop learning the old languages (as is at present threatening) a new class of literature will spring up, consisting of writing that is more barbaric, stupid, and worthless than has ever yet existed; that, in particular, the German language, which possesses some of the beauties of the old languages, will be systematically spoilt and stripped by these worthless contemporary scribblers, until, little by little, it becomes impoverished, crippled, and reduced to a miserable jargon.

Half a century is always a considerable time in the history of the universe, for the matter which forms it is always shifting; something is always taking place. But the same length of time in literature often goes for nothing, because nothing has happened; unskilful attempts don't count; so that we are exactly where we were fifty years previously.

To illustrate this: imagine the progress of knowledge among mankind in the form of a planet's course. The false paths the human race soon follows after any important progress has been made represent the epicycles in the Ptolemaic system; after passing through any one of them the planet is just where it was before it entered it. The great minds, however, which really bring the race further on its course, do not accompany it on the epicycles which it makes every time. This explains why posthumous fame is got at the expense of contemporary fame, and *vice versa*. We have an instance of such an epicycle in the philosophy of Fichte and Schelling, crowned by Hegel's caricature of it. This epicycle issued from the limit to which philosophy had been finally brought by Kant, where I myself took it up again later to carry it further. In the interim the false philosophers I have mentioned, and some others, passed through their epicycle, which has just been terminated; hence the people who accompanied them are conscious of being exactly at the point from which they started.

This condition of things shows why the scientific, literary, and artistic spirit of the age is declared bankrupt about every thirty years. During that period the errors have increased to such an extent that they fall under the weight of their absurdity; while at the same time the opposition to them has become stronger. At this point there is a crash, which is followed by an error in the opposite direction. To show the course that is taken in its periodical return would be the true practical subject of the history of literature; little notice is taken of it, however. Moreover, through the comparative shortness of such periods, the data of remote times are with difficulty collected; hence the matter can be most conveniently observed in one's own age. An example of this taken from physical science is found in Werter's Neptunian geology. But let me keep to the example already quoted above, for it is nearest to us. In German philosophy Kant's brilliant period was immediately followed by another period, which aimed at being imposing rather than convincing. Instead of being solid and clear, it aimed at being brilliant and hyperbolical, and, in particular, unintelligible; instead of seeking truth, it intrigued. Under these circumstances philosophy could make no progress. Ultimately the whole school and its method became bankrupt. For the audacious, sophisticated nonsense on the one hand, and the unconscionable praise on the other of Hegel and his fellows, as well as the apparent object of the whole affair, rose to such a pitch that in the end the charlatanry of the thing was obvious to everybody; and when, in consequence of certain revelations, the protection that had been given it by the upper classes was withdrawn, it was talked about by everybody. This most miserable of all the philosophies that have ever existed dragged down with it into the abyss of discredit the systems of Fichte and Schelling, which had preceded it. So that the absolute philosophical futility of the first half of the century following upon Kant in Germany is obvious; and yet the Germans boast of their gift for philosophy compared with foreigners, especially since an English writer, with malicious irony, called them *a nation of thinkers.*

Those who want an example of the general scheme of epicycles taken from the history of art need only look at the School of Sculpture which flourished in the last century under Bernini, and especially at its further cultivation in France. This school represented commonplace nature instead of antique beauty, and the manners of a French minuet instead of antique simplicity and grace. It became bankrupt when, under Winckelmann's direction, a return was made to the antique school. Another example is supplied in the painting belonging to the first quarter of this century. Art was regarded merely as a means and instrument of mediaeval religious feeling, and consequently ecclesiastical subjects alone were chosen for its themes. These, however, were treated by painters who were wanting in earnestness of faith, and in their delusion they took for examples Francesco Francia, Pietro Perugino, Angelico da Fiesole, and others like them, even holding them in greater esteem than the truly great masters who followed. In view of this error, and because in poetry an analogous effort had at the same time met with favour, Goethe wrote his parable *Pfaffenspiel.* This school, reputedly capricious, became bankrupt, and was followed by a return to nature, which made itself known in *genre* pictures and scenes of life of every description, even though it strayed sometimes into vulgarity.

It is the same with the progress of the human mind in the *history of literature*, which is for the most part like the catalogue of a cabinet of deformities; the spirit in which they keep the longest is pigskin. We do not need to look there for the few who have been born shapely; they are still alive, and we come across them in every part of the world, like immortals whose youth is ever fresh. They alone form what I have distinguished as *real* literature, the history of which, although poor in persons, we learn from our youth up out of the mouths of educated people, and not first of all from compilations. As a specific against the present prevailing monomania for reading literary histories, so that one may be able to chatter about everything without really knowing anything, let me refer you to a passage from Lichtenberg which is well worth reading (vol. ii. p. 302 of the old edition).

But I wish some one would attempt a *tragical history of literature*, showing how the greatest writers and artists have been treated during their lives by the various nations which have produced them and whose proudest possessions they are. It would show us the endless fight which the good and genuine works of all periods and countries have had to carry on against the perverse and bad. It would depict the martyrdom of almost all those who truly enlightened humanity, of almost all the great masters in every kind of art; it would show us how they, with few exceptions, were tormented without recognition, without any to share their misery, without followers; how they existed in poverty and misery whilst fame, honour, and riches fell to the lot of the worthless; it would reveal that what happened to them happened to Esau, who, while hunting the deer for his father, was robbed of the blessing by Jacob disguised in his brother's coat; and how through it all the love of their subject kept them up, until at last the trying fight of such a teacher of the human race is ended, the immortal laurel offered to him, and the time come when it can be said of him

> "Der schwere Panzer wird zum Fluegelkleide
> Kurz ist der Schmerz, unendlich ist die Freude."

THE EMPTINESS OF EXISTENCE.

This emptiness finds its expression in the whole form of existence, in the infiniteness of Time and Space as opposed to the finiteness of the individual in both; in the flitting present as the only manner of real existence; in the dependence and relativity of all things; in constantly Becoming without Being; in continually wishing without being satisfied; in an incessant thwarting of one's efforts, which go to make up life, until victory is won. *Time*, and the *transitoriness* of all things, are merely the form under which the will to live, which as the thing-in-itself is imperishable, has revealed to Time the futility of its efforts. Time is that by which at every moment all things become as nothing in our hands, and thereby lose all their true value.

* * * * *

What *has been* exists no more; and exists just as little as that which has *never* been. But everything that exists *has been* in the next moment. Hence something belonging to the present, however unimportant it may be, is superior to something important belonging to the past; this is because the former is a *reality* and related to the latter as something is to nothing.

A man to his astonishment all at once becomes conscious of existing after having been in a state of non-existence for many thousands of years, when, presently again, he returns to a state of non-existence for an equally long time. This cannot possibly be true, says the heart; and even the crude mind, after giving the matter its consideration, must have some sort of presentiment of the ideality of time. This ideality of time, together with that of space, is the key to every true system of metaphysics, because it finds room for quite another order of things than is to be found in nature. This is why Kant is so great.

Of every event in our life it is only for a moment that we can say that it *is*; after that we must say for ever that it *was*. Every evening makes us poorer by a day. It would probably make us angry to see this short space of time slipping away, if we were not secretly conscious in the furthest depths of our being that the spring of eternity belongs to us, and that in it we are always able to have life renewed.

Reflections of the nature of those above may, indeed, establish the belief that to enjoy the present, and to make this the purpose of one's life, is the greatest *wisdom*; since it is the present alone that is real, everything else being only the play of thought. But such a purpose might just as well be called the greatest *folly*, for that which in the next moment exists no more, and vanishes as completely as a dream, can never be worth a serious effort.

* * * * *

Our existence is based solely on the ever-fleeting present. Essentially, therefore, it has to take the form of continual motion without there ever being any possibility of our finding the rest after which we are always striving. It is the same as a man running downhill, who falls if he tries to stop, and it is only by his continuing to run on that he keeps on his legs; it is like a pole balanced on one's finger-tips, or like a planet that would fall into its sun as soon as it stopped hurrying onwards. Hence unrest is the type of existence.

In a world like this, where there is no kind of stability, no possibility of anything lasting, but where everything is thrown into a restless whirlpool of change, where everything hurries on, flies, and is maintained in the balance by a continual advancing and moving, it is impossible to imagine happiness. It cannot dwell where, as Plato says, *continual Becoming and never*

73

Being is all that takes place. First of all, no man is happy; he strives his whole life long after imaginary happiness, which he seldom attains, and if he does, then it is only to be disillusioned; and as a rule he is shipwrecked in the end and enters the harbour dismasted. Then it is all the same whether he has been happy or unhappy in a life which was made up of a merely ever-changing present and is now at an end.

Meanwhile it surprises one to find, both in the world of human beings and in that of animals, that this great, manifold, and restless motion is sustained and kept going by the medium of two simple impulses—hunger and the instinct of sex, helped perhaps a little by boredom—and that these have the power to form the *primum mobile* of so complex a machinery, setting in motion the variegated show!

Looking at the matter a little closer, we see at the very outset that the existence of inorganic matter is being constantly attacked by chemical forces which eventually annihilates it. While organic existence is only made possible by continual change of matter, to keep up a perpetual supply of which it must consequently have help from without. Therefore organic life is like balancing a pole on one's hand; it must be kept in continual motion, and have a constant supply of matter of which it is continually and endlessly in need. Nevertheless it is only by means of this organic life that consciousness is possible.

Accordingly this is a *finite existence*, and its antithesis would be an *infinite*, neither exposed to any attack from without nor in want of help from without, and hence [Greek: aei hosautos on], in eternal rest; [Greek: oute gignomenon, oute apollymenon], without change, without time, and without diversity; the negative knowledge of which is the fundamental note of Plato's philosophy. The denial of the will to live reveals the way to such a state as this.

* * * * *

The scenes of our life are like pictures in rough mosaic, which have no effect at close quarters, but must be looked at from a distance in order to discern their beauty. So that to obtain something we have desired is to find out that it is worthless; we are always living in expectation of better things, while, at the same time, we often repent and long for things that belong to the past. We accept the present as something that is only temporary, and regard it only as a means to accomplish our aim. So that most people will find if they look back when their life is at an end, that they have lived their lifelong *ad interim*, and they will be surprised to find that something they allowed to pass by unnoticed and unenjoyed was just their life—that is to say, it was the very thing in the expectation of which they lived. And so it may be said of man in general that, befooled by hope, he dances into the arms of death.

Then again, there is the insatiability of each individual will; every time it is satisfied a new wish is engendered, and there is no end to its eternally insatiable desires.

This is because the Will, taken in itself, is the lord of worlds; since everything belongs to it, it is not satisfied with a portion of anything, but only with the whole, which, however, is endless. Meanwhile it must excite our pity when we consider how extremely little this lord of the world receives, when it makes its appearance as an individual; for the most part only just enough to maintain the body. This is why man is so very unhappy.

In the present age, which is intellectually impotent and remarkable for its veneration of what is bad in every form—a condition of things which is quite in keeping with the coined word "Jetztzeit" (present time), as pretentious as it is cacophonic—the pantheists make bold to say that life is, as they call it, "an end-in itself." If our existence in this world were an end-in-itself, it would be the most absurd end that was ever determined; even we ourselves or any one else might have imagined it.

Life presents itself next as a task, the task, that is, of subsisting *de gagner sa vie*. If this is solved, then that which has been won becomes a burden, and involves the second task of its being got rid of in order to ward off boredom, which, like a bird of prey, is ready to fall upon any life that is secure from want.

So that the first task is to win something, and the second, after the something has been won, to forget about it, otherwise it becomes a burden.

That human life must be a kind of mistake is sufficiently clear from the fact that man is a compound of needs, which are difficult to satisfy; moreover, if they are satisfied, all he is granted is a state of painlessness, in which he can only give himself up to boredom. This is a precise proof that existence in itself has no value, since boredom is merely the feeling of the emptiness of life. If, for instance, life, the longing for which constitutes our very being, had in itself any positive and real value, boredom could not exist; mere existence in itself would supply us with everything, and therefore satisfy us. But our existence would not be a joyous thing unless we were striving after something; distance and obstacles to be overcome then represent our aim as something that would satisfy us—an illusion which vanishes when our aim has been attained; or when we are engaged in something that is of a purely intellectual nature, when, in reality, we have retired from the world, so that we may observe it from the outside, like spectators at a theatre. Even sensual pleasure itself is nothing but a continual striving, which ceases directly its aim is attained. As soon as we are not engaged in one of these two ways, but thrown back on existence itself, we are convinced of the emptiness and worthlessness of it; and this it is we call boredom. That innate and ineradicable craving for what is out of the common proves how glad we are to have the natural and tedious course of things interrupted. Even the pomp and splendour of the rich in their stately castles is at bottom nothing but a futile attempt to escape the very essence of existence, *misery*.

* * * * *

That the most perfect manifestation of the *will to live*, which presents itself in the extremely subtle and complicated machinery of the human organism, must fall to dust and finally deliver up its whole being to dissolution, is the naive way in which Nature, invariably true and genuine, declares the whole striving of the will in its very essence to be of no avail. If it were of any value in itself, something unconditioned, its end would not be non-existence. This is the dominant note of Goethe's beautiful song:

> "Hoch auf dem alten Thurme steht
> Des Helden edler Geist."

That man is nothing but a phenomenon, that he is not-the-thing-in-itself—I mean that he is not [Greek: ontos on]—is proved by the fact that *death is a necessity*.

And how different the beginning of our life is to the end! The former is made up of deluded hopes, sensual enjoyment, while the latter is pursued by bodily decay and the odour of death.

The road dividing the two, as far as our well-being and enjoyment of life are concerned, is downhill; the dreaminess of childhood, the joyousness of youth, the troubles of middle age, the infirmity and frequent misery of old age, the agonies of our last illness, and finally the struggle with death—do all these not make one feel that existence is nothing but a mistake, the consequences of which are becoming gradually more and more obvious?

It would be wisest to regard life as a *desengano*, a delusion; that everything is intended to be so is sufficiently clear.

Our life is of a microscopical nature; it is an indivisible point which, drawn out by the powerful lenses of Time and Space, becomes considerably magnified.

Time is an element in our brain which by the means of duration gives a semblance of reality to the *absolutely empty existence* of things and ourselves.

How foolish it is for a man to regret and deplore his having made no use of past opportunities, which might have secured him this or that happiness or enjoyment! What is there left of them now? Only the ghost of a remembrance! And it is the same with everything that really falls to our lot. So that the *form of time* itself, and how much is reckoned on it, is a definite way of proving to us the vanity of all earthly enjoyment.

Our existence, as well as that of all animals, is not one that lasts, it is only temporary, merely an *existentia fluxa*, which may be compared to a water-mill in that it is constantly changing.

It is true that the *form* of the body lasts for a time, but only on condition that the matter is constantly changing, that the old matter is thrown off and new added. And it is the chief work of all living creatures to secure a constant supply of suitable matter. At the same time, they are conscious that their existence is so fashioned as to last only for a certain time, as has been said. This is why they attempt, when they are taking leave of life, to hand it over to some one else who will take their place. This attempt takes the form of the sexual instinct in self-consciousness, and in the consciousness of other things presents itself objectively—that is, in the form of genital instinct. This instinct may be compared to the threading of a string of pearls; one individual succeeding another as rapidly as the pearls on the thread. If we, in imagination, hasten on this succession, we shall see that the matter is constantly changing in the whole row just as it is changing in each pearl, while it retains the same form: we will then realise that we have only a quasi-existence. That it is only Ideas which exist, and the shadow-like nature of the thing corresponding to them, is the basis of Plato's teachings.

That we are nothing but *phenomena* as opposed to the thing-in-itself is confirmed, exemplified, and made clear by the fact that the *conditio sine qua non* of our existence is a continual flowing off and flowing to of matter which, as nourishment, is a constant need. So that we resemble such phenomena as smoke, fire, or a jet of water, all of which die out or stop directly there is no supply of matter. It may be said then that the *will to live* presents itself in the form of *pure phenomena* which end *in nothing*. This nothingness, however, together with the phenomena, remain within the boundary of the *will to live* and are based on it. I admit that this is somewhat obscure.

If we try to get a general view of humanity at a glance, we shall see everywhere a constant fighting and mighty struggling for life and existence; that mental and bodily strength is taxed to the utmost, and opposed by threatening and actual dangers and woes of every kind.

And if we consider the price that is paid for all this, existence, and life itself, it will be found that there has been an interval when existence was free from pain, an interval, however, which was immediately followed by boredom, and which in its turn was quickly terminated by fresh cravings.

That boredom is immediately followed by fresh needs is a fact which is also true of the cleverer order of animals, because life has *no true and genuine value* in itself, but is kept *in motion* merely through the medium of needs and illusion. As soon as there are no needs and illusion we become conscious of the absolute barrenness and emptiness of existence.

If one turns from contemplating the course of the world at large, and in particular from the ephemeral and mock existence of men as they follow each other in rapid succession, to the *detail* of *life*, how like a comedy it seems!

It impresses us in the same way as a drop of water, crowded with *infusoria*, seen through a microscope, or a little heap of cheese-mites that would otherwise be invisible. Their activity and struggling with each other in such little space amuse us greatly. And it is the same in the little span of life—great and earnest activity produces a comic effect.

No man has ever felt perfectly happy in the present; if he had it would have intoxicated him.

ON WOMEN.

These few words of Jouy, *Sans les femmes le commencement de notre vie seroit prive de secours, le milieu de plaisirs et la fin de consolation*, more exactly express, in my opinion, the true praise of woman than Schiller's poem, *Wuerde der Frauen*, which is the fruit of much careful thought and impressive because of its antithesis and use of contrast. The same thing is more pathetically expressed by Byron in *Sardanapalus*, Act i, Sc. 2:—

> "The very first
> Of human life must spring from woman's breast,
> Your first small words are taught you from her lips,
> Your first tears quench'd by her, and your last sighs
> Too often breathed out in a woman's hearing,
> When men have shrunk from the ignoble care
> Of watching the last hour of him who led them."

Both passages show the right point of view for the appreciation of women.

One need only look at a woman's shape to discover that she is not intended for either too much mental or too much physical work. She pays the debt of life not by what she does but by what she suffers—by the pains of child-bearing, care for the child, and by subjection to man, to whom she should be a patient and cheerful companion. The greatest sorrows and joys or great exhibition of strength are not assigned to her; her life should flow more quietly, more gently, and less obtrusively than man's, without her being essentially happier or unhappier.

* * * * *

Women are directly adapted to act as the nurses and educators of our early childhood, for the simple reason that they themselves are childish, foolish, and short-sighted—in a word, are big children all their lives, something intermediate between the child and the man, who is a man in the strict sense of the word. Consider how a young girl will toy day after day with a child, dance with it and sing to it; and then consider what a man, with the very best intentions in the world, could do in her place.

* * * * *

With girls, Nature has had in view what is called in a dramatic sense a "striking effect," for she endows them for a few years with a richness of beauty and a, fulness of charm at the expense of the rest of their lives; so that they may during these years ensnare the fantasy of a man to such a degree as to make him rush into taking the honourable care of them, in some kind of form, for a lifetime—a step which would not seem sufficiently justified if he only considered the matter. Accordingly, Nature has furnished woman, as she has the rest of her creatures, with the weapons and implements necessary for the protection of her existence and for just the length of time that they will be of service to her; so that Nature has proceeded here with her usual economy. Just as the female ant after coition loses her wings, which then become superfluous, nay, dangerous for breeding purposes, so for the most part does a woman lose her beauty after giving birth to one or two children; and probably for the same reasons.

Then again we find that young girls in their hearts regard their domestic or other affairs as secondary things, if not as a mere jest. Love, conquests, and all that these include, such as dressing, dancing, and so on, they give their serious attention.

* * * * *

The nobler and more perfect a thing is, the later and slower is it in reaching maturity. Man reaches the maturity of his reasoning and mental faculties scarcely before he is eight-and-twenty; woman when she is eighteen; but hers is reason of very narrow limitations. This is why women remain children all their lives, for they always see only what is near at hand, cling to the present, take the appearance of a thing for reality, and prefer trifling matters to the most important. It is by virtue of man's reasoning powers that he does not live in the present only, like the brute, but observes and ponders over the past and future; and from this spring discretion, care, and that anxiety which we so frequently notice in people. The advantages, as well as the disadvantages, that this entails, make woman, in consequence of her weaker reasoning powers, less of a partaker in them. Moreover, she is intellectually short-sighted, for although her intuitive understanding quickly perceives what is near to her, on the other hand her circle of vision is limited and does not embrace anything that is remote; hence everything that is absent or past, or in the future, affects women in a less degree than men. This is why they have greater inclination for extravagance, which sometimes borders on madness. Women in their hearts think that men are intended to earn money so that they may spend it, if possible during their husband's lifetime, but at any rate after his death.

As soon as he has given them his earnings on which to keep house they are strengthened in this belief. Although all this entails many disadvantages, yet it has this advantage—that a woman lives more in the present than a man, and that she enjoys it more keenly if it is at all bearable. This is the origin of that cheerfulness which is peculiar to woman and makes her fit to divert man, and in case of need, to console him when he is weighed down by cares. To consult women in matters of difficulty, as the Germans used to do in old times, is by no means a matter to be overlooked; for their way of grasping a thing is quite different from ours, chiefly because they like the shortest way to the point, and usually keep their attention fixed upon what lies nearest; while we, as a rule, see beyond it, for the simple reason that it lies under our nose; it then becomes necessary for us to be brought back to the thing in order to obtain a near and simple view. This is why women are more sober in their judgment than we, and why they see nothing more in things than is really there; while we, if our passions are roused, slightly exaggerate or add to our imagination.

It is because women's reasoning powers are weaker that they show more sympathy for the unfortunate than men, and consequently take a kindlier interest in them. On the other hand, women are inferior to men in matters of justice, honesty, and conscientiousness. Again, because their reasoning faculty is weak, things clearly visible and real, and belonging to the present, exercise a power over them which is rarely counteracted by abstract thoughts, fixed maxims, or firm resolutions, in general, by regard for the past and future or by consideration for what is absent and remote. Accordingly they have the first and principal qualities of virtue, but they lack the secondary qualities which are often a necessary instrument in developing it. Women may be compared in this respect to an organism that has a liver but no gall-bladder.[9] So that it will be found that the fundamental fault in the character of women is that they have no *"sense of justice."* This arises from their deficiency in the power of reasoning already referred to, and reflection, but is also partly due to the fact that Nature has not destined them, as the weaker sex, to be dependent on strength but on cunning; this is why they are instinctively crafty, and have an ineradicable tendency to lie. For as lions are furnished with claws and teeth, elephants with tusks, boars with fangs, bulls with horns, and the cuttlefish with its dark, inky fluid, so Nature has provided woman for her protection and defence with the faculty of dissimulation, and all the power which Nature has given to man in the form of bodily strength and reason has been conferred on woman in this form. Hence, dissimulation is innate in woman and almost as characteristic of the very stupid as of the clever. Accordingly, it is as natural for women to dissemble at every opportunity as it is for those animals to turn to their weapons when they are attacked; and they feel in doing so that in a certain measure they are only making use of their rights. Therefore a woman who is perfectly truthful and does not dissemble is perhaps an impossibility. This is why they see through dissimulation in others so easily; therefore it is not advisable to attempt it with them. From the fundamental defect that has been stated, and all that it involves, spring falseness, faithlessness, treachery, ungratefulness, and so on. In a court of justice women are more often found guilty of perjury than men. It is indeed to be generally questioned whether they should be allowed to take an oath at all. From time to time there are repeated cases everywhere of ladies, who want for nothing, secretly pocketing and taking away things from shop counters.

* * * * *

Nature has made it the calling of the young, strong, and handsome men to look after the propagation of the human race; so that the species may not degenerate. This is the firm will of Nature, and it finds its expression in the passions of women. This law surpasses all others in both age and power. Woe then to the man who sets up rights and interests in such a way as to make them stand in the way of it; for whatever he may do or say, they will, at the first significant onset, be unmercifully annihilated. For the secret, unformulated, nay, unconscious but innate moral of woman is: *We are justified in deceiving those who, because they care a little for us,—that is to say for the individual,—imagine they have obtained rights over the species. The constitution, and consequently the welfare of the species, have been put into our hands and entrusted to our care through the medium of the next generation which proceeds from us; let us fulfil our duties conscientiously.*

But women are by no means conscious of this leading principle *in abstracto,* they are only conscious of it *in concreto,* and have no other way of expressing it than in the manner in which they act when the opportunity arrives. So that their conscience does not trouble them so much as we imagine, for in the darkest depths of their hearts they are conscious that in violating their duty towards the individual they have all the better fulfilled

84

it towards the species, whose claim upon them is infinitely greater. (A fuller explanation of this matter may be found in vol. ii., ch. 44, in my chief work, *Die Welt als Wille und Vorstellung*.)

Because women in truth exist entirely for the propagation of the race, and their destiny ends here, they live more for the species than for the individual, and in their hearts take the affairs of the species more seriously than those of the individual. This gives to their whole being and character a certain frivolousness, and altogether a certain tendency which is fundamentally different from that of man; and this it is which develops that discord in married life which is so prevalent and almost the normal state.

It is natural for a feeling of mere indifference to exist between men, but between women it is actual enmity. This is due perhaps to the fact that *odium figulinum* in the case of men, is limited to their everyday affairs, but with women embraces the whole sex; since they have only one kind of business. Even when they meet in the street, they look at each other like Guelphs and Ghibellines. And it is quite evident when two women first make each other's acquaintance that they exhibit more constraint and dissimulation than two men placed in similar circumstances. This is why an exchange of compliments between two women is much more ridiculous than between two men. Further, while a man will, as a rule, address others, even those inferior to himself, with a certain feeling of consideration and humanity, it is unbearable to see how proudly and disdainfully a lady of rank will, for the most part, behave towards one who is in a lower rank (not employed in her service) when she speaks to her. This may be because differences of rank are much more precarious with women than with us, and consequently more quickly change their line of conduct and elevate them, or because while a hundred things must be weighed in our case, there is only one to be weighed in theirs, namely, with which man they have found favour; and again, because of the one-sided nature of their vocation they stand in closer relationship to each other than men do; and so it is they try to render prominent the differences of rank.

* * * * *

It is only the man whose intellect is clouded by his sexual instinct that could give that stunted, narrow-shouldered, broad-hipped, and short-legged race the name of *the fair sex*; for the entire beauty of the sex is based on this instinct. One would be more justified in calling them the *unaesthetic sex* than the beautiful. Neither for music, nor for poetry, nor for fine art have they any real or true sense and susceptibility, and it is mere mockery on their part, in their desire to please, if they affect any such thing.

This makes them incapable of taking a purely objective interest in anything, and the reason for it is, I fancy, as follows. A man strives to get *direct* mastery over things either by understanding them or by compulsion. But a woman is always and everywhere driven to *indirect* mastery, namely through a man; all her *direct* mastery being limited to him alone. Therefore it lies in woman's nature to look upon everything only as a means for winning man, and her interest in anything else is always a simulated one, a mere roundabout way to gain her ends, consisting of coquetry and pretence. Hence Rousseau said, *Les femmes, en général, n'aiment aucun art, ne se connoissent a aucun et n'ont aucun genie* (Lettre a d'Alembert, note xx.). Every one who can see through a sham must have found this to be the case. One need only watch the way they behave at a concert, the opera, or the play; the childish simplicity, for instance, with which they keep on chattering during the finest passages in the greatest masterpieces. If it is true that the Greeks forbade women to go to the play, they acted in a right way; for they would at any rate be able to hear something. In our day it would be more appropriate to substitute *taceat mulier in theatro* for *taceat mulier in ecclesia*; and this might perhaps be put up in big letters on the curtain.

86

Nothing different can be expected of women if it is borne in mind that the most eminent of the whole sex have never accomplished anything in the fine arts that is really great, genuine, and original, or given to the world any kind of work of permanent value. This is most striking in regard to painting, the technique of which is as much within their reach as within ours; this is why they pursue it so industriously. Still, they have not a single great painting to show, for the simple reason that they lack that objectivity of mind which is precisely what is so directly necessary in painting. They always stick to what is subjective. For this reason, ordinary women have no susceptibility for painting at all: for *natura non facet saltum*. And Huarte, in his book which has been famous for three hundred years, *Examen de ingenios para las scienzias*, contends that women do not possess the higher capacities. Individual and partial exceptions do not alter the matter; women are and remain, taken altogether, the most thorough and incurable philistines; and because of the extremely absurd arrangement which allows them to share the position and title of their husbands they are a constant stimulus to his *ignoble* ambitions. And further, it is because they are philistines that modern society, to which they give the tone and where they have sway, has become corrupted. As regards their position, one should be guided by Napoleon's maxim, *Les femmes n'ont pas de rang*; and regarding them in other things, Chamfort says very truly: *Elles sont faites pour commercer avec nos faiblesses avec notre folie, mais non avec notre raison. Il existe entre elles et les hommes des sympathies d'epiderme et tres-peu de sympathies d'esprit d'ame et de caractere.* They are the *sexus sequior*, the second sex in every respect, therefore their weaknesses should be spared, but to treat women with extreme reverence is ridiculous, and lowers us in their own eyes. When nature divided the human race into two parts, she did not cut it exactly through the middle! The difference between the positive and negative poles, according to polarity, is not merely qualitative but also quantitative. And it was in this light that the ancients and people of the East regarded woman; they recognised her true position better than we, with our old French ideas of gallantry and absurd veneration, that highest product of Christian-Teutonic stupidity. These ideas have only served to make them arrogant and imperious, to such an extent as to remind one at times of the holy apes in Benares, who, in the consciousness of their holiness and inviolability, think they can do anything and everything they please.

In the West, the woman, that is to say the "lady," finds herself in a *fausse position*; for woman, rightly named by the ancients *sexus sequior*, is by no means fit to be the object of our honour and veneration, or to hold her head higher than man and to have the same rights as he. The consequences of this *fausse position* are sufficiently clear. Accordingly, it would be a very desirable thing if this Number Two of the human race in Europe were assigned her natural position, and the lady-grievance got rid of, which is not only ridiculed by the whole of Asia, but would have been equally ridiculed by Greece and Rome. The result of this would be that the condition of our social, civil, and political affairs would be incalculably improved. The Salic law would be unnecessary; it would be a superfluous truism. The European lady, strictly speaking, is a creature who should not exist at all; but there ought to be housekeepers, and young girls who hope to become such; and they should be brought up not to be arrogant, but to be domesticated and submissive. It is exactly because there are *ladies* in Europe that women of a lower standing, that is to say, the greater majority of the sex, are much more unhappy than they are in the East. Even Lord Byron says *(Letters and Papers*, by Thomas Moore, vol. ii. p. 399), *Thought of the state of women under the ancient Greeks—convenient enough. Present state, a remnant of the barbarism of the chivalric and feudal ages—artificial and unnatural. They ought to mind home—and be well fed and clothed—but not mixed in society. Well educated, too, in religion—but to read neither poetry nor politics—nothing but books of piety and cookery. Music—drawing—dancing—also a little gardening and ploughing now and then. I have seen them mending the roads in Epirus with good success. Why not, as well as hay-making and milking?*

* * * * *

In our part of the world, where monogamy is in force, to marry means to halve one's rights and to double one's duties. When the laws granted woman the same rights as man, they should also have given her a masculine power of reason. On the contrary, just as the privileges and honours which the laws decree to women surpass what Nature has meted out to them, so is there a proportional decrease in the number of women who really share these privileges; therefore the remainder are deprived of their natural rights in so far as the others have been given more than Nature accords.

For the unnatural position of privilege which the institution of monogamy, and the laws of marriage which accompany it, assign to the woman, whereby she is regarded throughout as a full equivalent of the man, which she is not by any means, cause intelligent and prudent men to reflect a great deal before they make so great a sacrifice and consent to so unfair an arrangement. Therefore, whilst among polygamous nations every woman finds maintenance, where monogamy exists the number of married women is limited, and a countless number of women who are without support remain over; those in the upper classes vegetate as useless old maids, those in the lower are reduced to very hard work of a distasteful nature, or become prostitutes, and lead a life which is as joyless as it is void of honour. But under such circumstances they become a necessity to the masculine sex; so that their position is openly recognised as a special means for protecting from seduction those other women favoured by fate either to have found husbands, or who hope to find them. In London alone there are 80,000 prostitutes. Then what are these women who have come too quickly to this most terrible end but human sacrifices on the altar of monogamy? The women here referred to and who are placed in this wretched position are the inevitable counterbalance to the European lady, with her pretensions and arrogance. Hence polygamy is a real benefit to the female sex, taking it *as a whole*. And, on the other hand, there is no reason why a man whose wife suffers from chronic illness, or remains barren, or has gradually become too old for him, should not take a second. Many people become converts to Mormonism for the precise reasons that they condemn the unnatural institution of monogamy. The conferring of unnatural rights upon women has imposed unnatural duties upon them, the violation of which, however, makes them unhappy. For example, many a man thinks marriage unadvisable as far as his social standing and monetary position are concerned, unless he contracts a brilliant match. He will then wish to win a woman of his own choice under different conditions, namely, under those which will render safe her future and that of her children. Be the conditions ever so just, reasonable, and adequate, and she consents by giving up those undue privileges which marriage, as the basis of civil society, alone can bestow, she must to a certain extent lose her honour and lead a life of loneliness; since human nature makes us dependent on the opinion of others in a way that is completely out of proportion to its value. While, if the woman does not consent, she runs the risk of being compelled to marry a man she dislikes, or of shrivelling up into an old maid; for the time allotted to her to find a home is very short. In view of this side of the institution of monogamy, Thomasius's profoundly learned treatise, *de Concubinatu*, is well worth reading, for it shows that, among all nations, and in all ages, down to the Lutheran Reformation, concubinage was allowed, nay, that it was an institution, in a certain measure even recognised by law and associated with no dishonour. And it held this position until the Lutheran Reformation, when it was recognised as another means for justifying the marriage of the clergy; whereupon the Catholic party did not dare to remain behindhand in the matter.

It is useless to argue about polygamy, it must be taken as a fact existing everywhere, the *mere regulation* of which is the problem to be solved. Where are there, then, any real monogamists? We all live, at any rate for a time, and the majority of us always, in polygamy. Consequently, as each man needs many women, nothing is more just than to let him, nay, make it incumbent upon him to provide for many women. By this means woman will be brought back to her proper and natural place as a subordinate being, and *the lady*, that monster of European civilisation and Christian-Teutonic stupidity, with her ridiculous claim to respect and veneration, will no longer exist; there will still be *women*, but no *unhappy women*, of whom Europe is at present full. The Mormons' standpoint is right.

* * * * *

In India no woman is ever independent, but each one stands under the control of her father or her husband, or brother or son, in accordance with the law of Manu.

It is certainly a revolting idea that widows should sacrifice themselves on their husband's dead body; but it is also revolting that the money which the husband has earned by working diligently for all his life, in the hope that he was working for his children, should be wasted on her paramours. *Medium tenuere beati.* The first love of a mother, as that of animals and men, is purely *instinctive*, and consequently ceases when the child is no longer physically helpless. After that, the first love should be reinstated by a love based on habit and reason; but this often does not appear, especially where the mother has not loved the father. The love of a father for his children is of a different nature and more sincere; it is founded on a recognition of his own inner self in the child, and is therefore metaphysical in its origin.

In almost every nation, both of the new and old world, and even among the Hottentots, property is inherited by the male descendants alone; it is only in Europe that one has departed from this. That the property which men have with difficulty acquired by long-continued struggling and hard work should afterwards come into the hands of women, who, in their want of reason, either squander it within a short time or otherwise waste it, is an injustice as great as it is common, and it should be prevented by limiting the right of women to inherit. It seems to me that it would be a better arrangement if women, be they widows or daughters, only inherited the money for life secured by mortgage, but not the property itself or the capital, unless there lacked male descendants. It is men who make the money, and not women; therefore women are neither justified in having unconditional possession of it nor capable of administrating it. Women should never have the free disposition of wealth, strictly so-called, which they may inherit, such as capital, houses, and estates. They need a guardian always; therefore they should not have the guardianship of their children under any circumstances whatever. The vanity of women, even if it should not be greater than that of men, has this evil in it, that it is directed on material things—that is to say, on their personal beauty and then on tinsel, pomp, and show. This is why they are in their right element in society. This it is which makes them inclined to be *extravagant*, especially since they possess little reasoning power. Accordingly, an ancient writer says, [Greek: Gunae to synolon esti dapanaeron physei].[10] Men's vanity, on the other hand, is often directed on non-material advantages, such as intellect, learning, courage, and the like. Aristotle explains in the *Politics*[11] the great disadvantages which the Spartans brought upon themselves by granting too much to their women, by allowing them the right of inheritance and dowry, and a great amount of freedom; and how this contributed greatly to the fall of Sparta. May it not be that the influence of women in France, which has been increasing since Louis XIII.'s time, was to blame for that gradual corruption of the court and government which led to the first Revolution, of which all subsequent disturbances have been the result? In any case, the false position of the female sex, so conspicuously exposed by the existence of the "lady," is a fundamental defect in our social condition, and this defect, proceeding from the very heart of it, must extend its harmful influence in every direction. That woman is by nature intended to obey is shown by the fact that every woman who is placed in the unnatural position of absolute independence at once attaches herself to some kind of man, by whom she is controlled and governed; this is because she requires a master. If she, is young, the man is a lover; if she is old, a priest.

FOOTNOTES:

9   Let me refer to what I have said in my treatise on *The Foundation of Morals,* Sec.71.

10   Brunck's *Gnomici poetae graeci* v. 115.

11   Bk. I., ch. 9.

THINKING FOR ONESELF.

The largest library in disorder is not so useful as a smaller but orderly one; in the same way the greatest amount of knowledge, if it has not been worked out in one's own mind, is of less value than a much smaller amount that has been fully considered. For it is only when a man combines what he knows from all sides, and compares one truth with another, that he completely realises his own knowledge and gets it into his power. A man can only think over what he knows, therefore he should learn something; but a man only knows what he has pondered.

A man can apply himself of his own free will to reading and learning, while he cannot to thinking. Thinking must be kindled like a fire by a draught and sustained by some kind of interest in the subject. This interest may be either of a purely objective nature or it may be merely subjective. The latter exists in matters concerning us personally, but objective interest is only to be found in heads that think by nature, and to whom thinking is as natural as breathing; but they are very rare. This is why there is so little of it in most men of learning.

The difference between the effect that thinking for oneself and that reading has on the mind is incredibly great; hence it is continually developing that original difference in minds which induces one man to think and another to read. Reading forces thoughts upon the mind which are as foreign and heterogeneous to the bent and mood in which it may be for the moment, as the seal is to the wax on which it stamps its imprint. The mind thus suffers total compulsion from without; it has first this and first that to think about, for which it has at the time neither instinct nor liking.

On the other hand, when a man thinks for himself he follows his own impulse, which either his external surroundings or some kind of recollection has determined at the moment. His visible surroundings do not leave upon his mind *one* single definite thought as reading does, but merely supply him with material and occasion to think over what is in keeping with his nature and present mood. This is why *much* reading robs the mind of all elasticity; it is like keeping a spring under a continuous, heavy weight. If a man does not want to think, the safest plan is to take up a book directly he has a spare moment.

This practice accounts for the fact that learning makes most men more stupid and foolish than they are by nature, and prevents their writings from being a success; they remain, as Pope has said,

"For ever reading, never to be read."

—*Dunciad* iii. 194.

Men of learning are those who have read the contents of books. Thinkers, geniuses, and those who have enlightened the world and furthered the race of men, are those who have made direct use of the book of the world.

* * * * *

Indeed, it is only a man's own fundamental thoughts that have truth and life in them. For it is these that he really and completely understands. To read the thoughts of others is like taking the remains of some one else's meal, like putting on the discarded clothes of a stranger.

The thought we read is related to the thought which rises in us, as the fossilised impress of a prehistoric plant is to a plant budding out in spring.

* * * * *

Reading is merely a substitute for one's own thoughts. A man allows his thoughts to be put into leading-strings.

Further, many books serve only to show how many wrong paths there are, and how widely a man may stray if he allows himself to be led by them. But he who is guided by his genius, that is to say, he who thinks for himself, who thinks voluntarily and rightly, possesses the compass wherewith to find the right course. A man, therefore, should only read when the source of his own thoughts stagnates; which is often the case with the best of minds.

It is sin against the Holy Spirit to frighten away one's own original thoughts by taking up a book. It is the same as a man flying from Nature to look at a museum of dried plants, or to study a beautiful landscape in copperplate. A man at times arrives at a truth or an idea after spending much time in thinking it out for himself, linking together his various thoughts, when he might have found the same thing in a book; it is a hundred times more valuable if he has acquired it by thinking it out for himself. For it is only by his thinking it out for himself that it enters as an integral part, as a living member into the whole system of his thought, and stands in complete and firm relation with it; that it is fundamentally understood with all its consequences, and carries the colour, the shade, the impress of his own way of thinking; and comes at the very moment, just as the necessity for it is felt, and stands fast and cannot be forgotten. This is the perfect application, nay, interpretation of Goethe's

> "Was du ererbt von deinen Vaetern hast
> Erwirb es um es zu besitzen."

The man who thinks for himself learns the authorities for his opinions only later on, when they serve merely to strengthen both them and himself; while the book-philosopher starts from the authorities and other people's opinions, therefrom constructing a whole for himself; so that he resembles an automaton, whose composition we do not understand. The other man, the man who thinks for himself, on the other hand, is like a living man as made by nature. His mind is impregnated from without, which then bears and brings forth its child. Truth that has been merely learned adheres to us like an artificial limb, a false tooth, a waxen nose, or at best like one made out of another's flesh; truth which is acquired by thinking for oneself is like a natural member: it alone really belongs to us. Here we touch upon the difference between the thinking man and the mere man of learning. Therefore the intellectual acquirements of the man who thinks for himself are like a fine painting that stands out full of life, that has its light and shade correct, the tone sustained, and perfect harmony of colour. The intellectual attainments of the merely learned man, on the contrary, resemble a big palette covered with every colour, at most systematically arranged, but without harmony, relation, and meaning.

*　*　*　*　*

Reading is thinking with some one else's head instead of one's own. But to think for oneself is to endeavour to develop a coherent whole, a system, even if it is not a strictly complete one. Nothing is more harmful than, by dint of continual reading, to strengthen the current of other people's thoughts. These thoughts, springing from different minds, belonging to different systems, bearing different colours, never flow together of themselves into a unity of thought, knowledge, insight, or conviction, but rather cram the head with a Babylonian confusion of tongues; consequently the mind becomes overcharged with them and is deprived of all clear insight and almost disorganised. This condition of things may often be discerned in many men of learning, and it makes them inferior in sound understanding, correct judgment, and practical tact to many illiterate men, who, by the aid of experience, conversation, and a little reading, have acquired a little knowledge from without, and made it always subordinate to and incorporated it with their own thoughts.

The scientific *thinker* also does this to a much greater extent. Although he requires much knowledge and must read a great deal, his mind is nevertheless strong enough to overcome it all, to assimilate it, to incorporate it with the system of his thoughts, and to subordinate it to the organic relative unity of his insight, which is vast and ever-growing. By this means his own thought, like the bass in an organ, always takes the lead in everything, and is never deadened by other sounds, as is the case with purely antiquarian minds; where all sorts of musical passages, as it were, run into each other, and the fundamental tone is entirely lost.

* * * * *

The people who have spent their lives in reading and acquired their wisdom out of books resemble those who have acquired exact information of a country from the descriptions of many travellers. These people can relate a great deal about many things; but at heart they have no connected, clear, sound knowledge of the condition of the country. While those who have spent their life in thinking are like the people who have been to that country themselves; they alone really know what it is they are saying, know the subject in its entirety, and are quite at home in it.

* * * * *

The ordinary book-philosopher stands in the same relation to a man who thinks for himself as an eye-witness does to the historian; he speaks from his own direct comprehension of the subject.

Therefore all who think for themselves hold at bottom much the same views; when they differ it is because they hold different points of view, but when these do not alter the matter they all say the same thing. They merely express what they have grasped from an objective point of view. I have frequently hesitated to give passages to the public because of their paradoxical nature, and afterwards to my joyful surprise have found the same thoughts expressed in the works of great men of long ago.

98

The book-philosopher, on the other hand, relates what one man has said and another man meant, and what a third has objected to, and so on. He compares, weighs, criticises, and endeavours to get at the truth of the thing, and in this way resembles the critical historian. For instance, he will try to find out whether Leibnitz was not for some time in his life a follower of Spinoza, etc. The curious student will find striking examples of what I mean in Herbart's *Analytical Elucidation of Morality and Natural Right*, and in his *Letters on Freedom*. It surprises us that such a man should give himself so much trouble; for it is evident that if he had fixed his attention on the matter he would soon have attained his object by thinking a little for himself.

But there is a small difficulty to overcome; a thing of this kind does not depend upon our own will. One can sit down at any time and read, but not—think. It is with thoughts as with men: we cannot always summon them at pleasure, but must wait until they come. Thought about a subject must come of its own accord by a happy and harmonious union of external motive with mental temper and application; and it is precisely that which never seems to come to these people.

One has an illustration of this in matters that concern our personal interest. If we have to come to a decision on a thing of this kind we cannot sit down at any particular moment and thrash out the reasons and arrive at a decision; for often at such a time our thoughts cannot be fixed, but will wander off to other things; a dislike to the subject is sometimes responsible for this. We should not use force, but wait until the mood appears of itself; it frequently comes unexpectedly and even repeats itself; the different moods which possess us at the different times throwing another light on the matter. It is this long process which is understood by *a ripe resolution*. For the task of making up our mind must be distributed; much that has been previously overlooked occurs to us; the aversion also disappears, for, after examining the matter closer, it seems much more tolerable than it was at first sight.

And in theory it is just the same: a man must wait for the right moment; even the greatest mind is not always able to think for itself at all times. Therefore it is advisable for it to use its spare moments in reading, which, as has been said, is a substitute for one's own thought; in this way material is imported to the mind by letting another think for us, although it is always in a way which is different from our own. For this reason a man should not read too much, in order that his mind does not become accustomed to the substitute, and consequently even forget the matter in question; that it may not get used to walking in paths that have already been trodden, and by following a foreign course of thought forget its own. Least of all should a man for the sake of reading entirely withdraw his attention from the real world: as the impulse and temper which lead one to think for oneself proceed oftener from it than from reading; for it is the visible and real world in its primitiveness and strength that is the natural subject of the thinking mind, and is able more easily than anything else to rouse it. After these considerations it will not surprise us to find that the thinking man can easily be distinguished from the book-philosopher by his marked earnestness, directness, and originality, the personal conviction of all his thoughts and expressions: the book-philosopher, on the other hand, has everything second-hand; his ideas are like a collection of old rags obtained anyhow; he is dull and pointless, resembling a copy of a copy. His style, which is full of conventional, nay, vulgar phrases and current terms, resembles a small state where there is a circulation of foreign money because it coins none of its own.

* * * * *

Mere experience can as little as reading take the place of thought. Mere empiricism bears the same relation to thinking as eating to digestion and assimilation. When experience boasts that it alone, by its discoveries, has advanced human knowledge, it is as though the mouth boasted that it was its work alone to maintain the body.

100

The works of all really capable minds are distinguished from all other works by a character of decision and definiteness, and, in consequence, of lucidity and clearness. This is because minds like these know definitely and clearly what they wish to express—whether it be in prose, in verse, or in music. Other minds are wanting in this decision and clearness, and therefore may be instantly recognised.

The characteristic sign of a mind of the highest standard is the directness of its judgment. Everything it utters is the result of thinking for itself; this is shown everywhere in the way it gives expression to its thoughts. Therefore it is, like a prince, an imperial director in the realm of intellect. All other minds are mere delegates, as may be seen by their style, which has no stamp of its own.

Hence every true thinker for himself is so far like a monarch; he is absolute, and recognises nobody above him. His judgments, like the decrees of a monarch, spring from his own sovereign power and proceed directly from himself. He takes as little notice of authority as a monarch does of a command; nothing is valid unless he has himself authorised it. On the other hand, those of vulgar minds, who are swayed by all kinds of current opinions, authorities, and prejudices, are like the people which in silence obey the law and commands.

* * * * *

The people who are so eager and impatient to settle disputed questions, by bringing forward authorities, are really glad when they can place the understanding and insight of some one else in the field in place of their own, which are deficient. Their number is legion. For, as Seneca says, *"Unusquisque mavult credere, quam judicare."*

The weapon they commonly use in their controversies is that of authorities: they strike each other with it, and whoever is drawn into the fray will do well not to defend himself with reason and arguments; for against a weapon of this kind they are like horned Siegfrieds, immersed in a flood of incapacity for thinking and judging. They will bring forward their authorities as an *argumentum ad verecundiam* and then cry *victoria*.

* * * * *

In the realm of reality, however fair, happy, and pleasant it may prove to be, we always move controlled by the law of gravity, which we must be unceasingly overcoming. While in the realm of thought we are disembodied spirits, uncontrolled by the law of gravity and free from penury.

This is why there is no happiness on earth like that which at the propitious moment a fine and fruitful mind finds in itself.

* * * * *

The presence of a thought is like the presence of our beloved. We imagine we shall never forget this thought, and that this loved one could never be indifferent to us. But out of sight out of mind! The finest thought runs the risk of being irrevocably forgotten if it is not written down, and the dear one of being forsaken if we do not marry her.

* * * * *

There are many thoughts which are valuable to the man who thinks them; but out of them only a few which possess strength to produce either repercussion or reflex action, that is, to win the reader's sympathy after they have been written down. It is what a man has thought out directly *for himself* that alone has true value. Thinkers may be classed as follows: those who, in the first place, think for themselves, and those who think directly for others. The former thinkers are the genuine, *they think for themselves* in both senses of the word; they are the true *philosophers*; they alone are in earnest. Moreover, the enjoyment and happiness of their existence consist in thinking. The others are the *sophists*; they wish to *seem*, and seek their happiness in what they hope to get from other people; their earnestness consists in this. To which of these two classes a man belongs is soon seen by his whole method and manner. Lichtenberg is an example of the first class, while Herder obviously belongs to the second.

* * * * *

When one considers how great and how close to us the *problem of existence* is,—this equivocal, tormented, fleeting, dream-like existence—so great and so close that as soon as one perceives it, it overshadows and conceals all other problems and aims;—and when one sees how all men—with a few and rare exceptions—are not clearly conscious of the problem, nay, do not even seem to see it, but trouble themselves about everything else rather than this, and live on taking thought only for the present day and the scarcely longer span of their own personal future, while they either expressly give the problem up or are ready to agree with it, by the aid of some system of popular metaphysics, and are satisfied with this;—when one, I say, reflects upon this, so may one be of the opinion that man is a *thinking being* only in a very remote sense, and not feel any special surprise at any trait of thoughtlessness or folly; but know, rather, that the intellectual outlook of the normal man indeed surpasses that of the brute,—whose whole existence resembles a continual present without any consciousness of the future or the past—but, however, not to such an extent as one is wont to suppose.

And corresponding to this, we find in the conversation of most men that their thoughts are cut up as small as chaff, making it impossible for them to spin out the thread of their discourse to any length. If this world were peopled by really thinking beings, noise of every kind would not be so universally tolerated, as indeed the most horrible and aimless form of it is.[12] If Nature had intended man to think she would not have given him ears, or, at any rate, she would have furnished them with air-tight flaps like the bat, which for this reason is to be envied. But, in truth, man is like the rest, a poor animal, whose powers are calculated only to maintain him during his existence; therefore he requires to have his ears always open to announce of themselves, by night as by day, the approach of the pursuer.

FOOTNOTES:

12   See Essay on Noise, p. 28.

SHORT DIALOGUE ON

THE INDESTRUCTIBILITY OF OUR TRUE BEING BY DEATH.

Thrasymachos. Tell me briefly, what shall I be after my death? Be clear and precise.

Philalethes. Everything and nothing.

Thras. That is what I expected. You solve the problem by a contradiction. That trick is played out.

Phil. To answer transcendental questions in language that is made for immanent knowledge must assuredly lead to a contradiction.

Thras. What do you call transcendental knowledge, and what immanent? It is true these expressions are known to me, for my professor used them, but only as predicates of God, and as his philosophy had exclusively to do with God, their use was quite appropriate. For instance, if God was in the world, He was immanent; if He was somewhere outside it, He was transcendent. That is clear and comprehensible. One knows how things stand. But your old-fashioned Kantian doctrine is no longer understood. There has been quite a succession of great men in the metropolis of German learning—

Phil. (aside). German philosophical nonsense!

Thras.—such as the eminent Schleiermacher and that gigantic mind Hegel; and to-day we have left all that sort of thing behind, or rather we are so far ahead of it that it is out of date and known no more. Therefore, what good is it?

105

Phil. Transcendental knowledge is that which, going beyond the boundary of possible experience, endeavours to determine the nature of things as they are in themselves; while immanent knowledge keeps itself within the boundary of possible experience, therefore it can only apply to phenomena. As an individual, with your death there will be an end of you. But your individuality is not your true and final being, indeed it is rather the mere expression of it; it is not the thing-in-itself but only the phenomenon presented in the form of time, and accordingly has both a beginning and an end. Your being in itself, on the contrary, knows neither time, nor beginning, nor end, nor the limits of a given individuality; hence no individuality can be without it, but it is there in each and all. So that, in the first sense, after death you become nothing; in the second, you are and remain everything. That is why I said that after death you would be all and nothing. It is difficult to give you a more exact answer to your question than this and to be brief at the same time; but here we have undoubtedly another contradiction; this is because your life is in time and your immortality in eternity. Hence your immortality may be said to be something that is indestructible and yet has no endurance—which is again contradictory, you see. This is what happens when transcendental knowledge is brought within the boundary of immanent knowledge; in doing this some sort of violence is done to the latter, since it is used for things for which it was not intended.

Thras. Listen; without I retain my individuality I shall not give a *sou* for your immortality.

Phil. Perhaps you will allow me to explain further. Suppose I guarantee that you will retain your individuality, on condition, however, that you spend three months in absolute unconsciousness before you awaken.

Thras. I consent to that.

106

Phil. Well then, as we have no idea of time when in a perfectly unconscious state, it is all the same to us when we are dead whether three months or ten thousand years pass away in the world of consciousness. For in the one case, as in the other, we must accept on faith and trust what we are told when we awake. Accordingly it will be all the same to you whether your individuality is restored to you after the lapse of three months or ten thousand years.

Thras. At bottom, that cannot very well be denied.

Phil. But if, at the end of those ten thousand years, some one has quite forgotten to waken you, I imagine that you would have become accustomed to that long state of non-existence, following such a very short existence, and that the misfortune would not be very great. However, it is quite certain that you would know nothing about it. And again, it would fully console you to know that the mysterious power which gives life to your present phenomenon had never ceased for one moment during the ten thousand years to produce other phenomena of a like nature and to give them life.

Thras. Indeed! And so it is in this way that you fancy you can quietly, and without my knowing, cheat me of my individuality? But you cannot cozen me in this way. I have stipulated for the retaining of my individuality, and neither mysterious forces nor phenomena can console me for the loss of it. It is dear to me, and I shall not let it go.

Phil. That is to say, you regard your individuality as something so very delightful, excellent, perfect, and incomparable that there is nothing better than it; would you not exchange it for another, according to what is told us, that is better and more lasting?

Thras. Look here, be my individuality what it may, it is myself,

"For God is God, and I am I."

I—I—I want to exist! That is what I care about, and not an existence which has to be reasoned out first in order to show that it is mine.

Phil. Look what you are doing! When you say, *I—I—I want to exist* you alone do not say this, but everything, absolutely everything, that has only a vestige of consciousness. Consequently this desire of yours is just that which is *not* individual but which is common to all without distinction. It does not proceed from individuality, but from *existence* in general; it is the essential in everything that exists, nay, it is *that* whereby anything has existence at all; accordingly it is concerned and satisfied only with existence *in general* and not with any definite individual existence; this is not its aim. It has the appearance of being so because it can attain consciousness only in an individual existence, and consequently looks as if it were entirely concerned with that. This is nothing but an illusion which has entangled the individual; but by reflection, it can be dissipated and we ourselves set free. It is only *indirectly* that the individual has this great longing for existence; it is the will to live in general that has this longing directly and really, a longing that is one and the same in everything. Since, then, existence itself is the free work of the will, nay, the mere reflection of it, existence cannot be apart from will, and the latter will be provisionally satisfied with existence in general, in so far, namely, as that which is eternally dissatisfied can be satisfied. The will is indifferent to individuality; it has nothing to do with it, although it appears to, because the individual is *only* directly conscious of will in himself. From this it is to be gathered that the individual carefully guards his own existence; moreover, if this were not so, the preservation of the species would not be assured. From all this it follows that individuality is not a state of perfection but of limitation; so that to be freed from it is not loss but rather gain. Don't let this trouble you any further, it will, forsooth, appear to you both childish and extremely ridiculous when you completely and thoroughly recognise what you are, namely, that your own existence is the universal will to live.

Thras. You are childish yourself and extremely ridiculous, and so are all philosophers; and when a sedate man like myself lets himself in for a quarter of an hour's talk with such fools, it is merely for the sake of amusement and to while away the time. I have more important matters to look to now; so, adieu!

RELIGION.

A DIALOGUE.

Demopheles. Between ourselves, dear old friend, I am sometimes dissatisfied
with you in your capacity as philosopher; you talk sarcastically about
religion, nay, openly ridicule it. The religion of every one is sacred
to him, and so it should be to you.

Philalethes. Nego consequentiam! I don't see at all why I should have respect
for lies and frauds because other people are stupid. I respect truth
everywhere, and it is precisely for that reason that I cannot respect
anything that is opposed to it. My maxim is, *Vigeat veritas, et pereat
mundus*, the same as the lawyer's *Fiat justitia, et pereat mundus*. Every
profession ought to have an analogous device.

Demop. Then that of the medical profession would be, *Fiant pilulae, et
pereat mundus*, which would be the easiest to carry out.

Phil. Heaven forbid! Everything must be taken *cum grano salis*.

Demop. Exactly; and it is just for that reason that I want you to accept
religion *cum grano salis*, and to see that the needs of the people must
be met according to their powers of comprehension. Religion
affords the only means of proclaiming and making the masses of
crude minds and awkward intelligences, sunk in petty pursuits and
material work, feel the high import of life. For the ordinary type of
man, primarily, has no thought for anything else but what satisfies
his physical needs and longings, and accordingly affords him a little
amusement and pastime. Founders of religion and philosophers

come into the world to shake him out of his torpidity and show him the high significance of existence: philosophers for the few, the emancipated; founders of religion for the many, humanity at large. For [Greek: philosophon plaethos adynaton einai], as your friend Plato has said, and you should not forget it. Religion is the metaphysics of the people, which by all means they must keep; and hence it must be eternally respected, for to discredit it means taking it away. Just as there is popular poetry, popular wisdom in proverbs, so too there must be popular metaphysics; for mankind requires most certainly *an interpretation of life*, and it must be in keeping with its power of comprehension. So that this interpretation is at all times an allegorical investiture of the truth, and it fulfils, as far as practical life and our feelings are concerned—that is to say, as a guidance in our affairs, and as a comfort and consolation in suffering and death—perhaps just as much as truth itself could, if we possessed it. Don't be hurt at its unpolished, baroque, and apparently absurd form, for you, with your education and learning, cannot imagine the roundabout ways that must be used in order to make people in their crude state understand deep truths. The various religions are only various forms in which the people grasp and understand the truth, which in itself they could not grasp, and which is inseparable from these forms. Therefore, my dear fellow, don't be displeased if I tell you that to ridicule these forms is both narrow-minded and unjust.

Phil. But is it not equally narrow-minded and unjust to require that there shall be no other metaphysics but this one cut out to meet the needs and comprehension of the people? that its teachings shall be the boundary of human researches and the standard of all thought, so that the metaphysics of the few, the emancipated, as you call them, must aim at confirming, strengthening, and interpreting the metaphysics of the people? That is, that the highest faculties of the human mind must remain unused and undeveloped, nay, be nipped in the bud, so that their activity may not thwart the popular metaphysics? And at bottom are not the claims that religion makes just the same? Is it right to have tolerance, nay, gentle forbearance, preached by what is intolerance and cruelty itself? Let me remind you of the heretical tribunals, inquisitions, religious wars and crusades, of Socrates' cup of poison, of Bruno's and Vanini's death in the flames. And is all this to-day something belonging to the past? What can stand more in the way of genuine philosophical effort, honest inquiry after truth, the noblest calling of the noblest of mankind, than this conventional system of metaphysics invested with a monopoly from the State, whose principles are inculcated so earnestly, deeply, and firmly into every head in earliest youth as to make them, unless the mind is of miraculous elasticity, become ineradicable? The result is that the basis of healthy reasoning is once and for all deranged—in other words, its feeble capacity for thinking for itself, and for unbiassed judgment in regard to everything to which it might be applied, is for ever paralysed and ruined.

Demop, Which really means that the people have gained a conviction which they will not give up in order to accept yours in its place.

Phil. Ah! if it were only conviction based on insight, one would then be able to bring forward arguments and fight the battle with equal weapons. But religions admittedly do not lend themselves to conviction after argument has been brought to bear, but to belief as brought about by revelation. The capacity for belief is strongest in childhood; therefore one is most careful to take possession of this tender age. It is much more through this than through threats and reports of miracles that the doctrines of belief take root. If in early childhood certain fundamental views and doctrines are preached with unusual solemnity and in a manner of great earnestness, the like of which has never been seen before, and if, too, the possibility of a doubt about them is either completely ignored or only touched upon in order to show that doubt is the first step to everlasting perdition; the result is that the impression will be so profound that, as a rule, that is to say in almost every case, a man will be almost as incapable of doubting the truth of those doctrines as he is of doubting his own existence. Hence it is scarcely one in many thousands that has the strength of mind to honestly and seriously ask himself—is that true? Those who are able to do this have been more appropriately styled strong minds, *esprits forts*, than is imagined. For the commonplace mind, however, there is nothing so absurd or revolting but what, if inoculated in this way, the firmest belief in it will take root. If, for example, the killing of a heretic or an infidel were an essential matter for the future salvation of the soul, almost every one would make it the principal object of his life, and in dying get consolation and strength from the remembrance of his having succeeded; just as, in truth, in former times almost every Spaniard looked upon an *auto da fe* as the most pious of acts and one most pleasing to God.

We have an analogy to this in India in the *Thugs*, a religious body quite recently suppressed by the English, who executed numbers of them. They showed their regard for religion and veneration for the goddess Kali by assassinating at every opportunity their own friends and fellow-travellers, so that they might obtain their possessions, and they were seriously convinced that thereby they had accomplished something that was praiseworthy and would contribute to their eternal welfare. The power of religious dogma, that has been inculcated early, is so great that it destroys conscience, and finally all compassion and sense of humanity. But if you wish to see with your own eyes, and close at hand, what early inoculation of belief does, look at the English. Look at this nation, favoured by nature before all others, endowed before all others with reason, intelligence, power of judgment, and firmness of character; look at these people degraded, nay, made despicable among all others by their stupid ecclesiastical superstition, which among their other capacities appears like a fixed idea, a monomania. For this they have to thank the clergy in whose hands education is, and who take care to inculcate all the articles, of belief at the earliest age in such a way as to result in a kind of partial paralysis of the brain; this then shows itself throughout their whole life in a silly bigotry, making even extremely intelligent and capable people among them degrade themselves so that they become quite an enigma to us. If we consider how essential to such a masterpiece is inoculation of belief in the tender age of childhood, the system of missions appears no longer merely as the height of human importunity, arrogance, and impertinence, but also of absurdity; in so far as it does not confine itself to people who are still in the stage of *childhood*, such as the Hottentots, Kaffirs, South Sea Islanders, and others like them, among whom it has been really successful. While, on the other hand, in India the Brahmans receive the doctrines of missionaries either with a smile of condescending approval or refuse them with a shrug of their shoulders; and among these people in general, notwithstanding the most favourable circumstances, the missionaries' attempts at conversion are usually wrecked. An authentic report in vol. xxi. of the *Asiatic Journal* of 1826 shows that after so many years of missionary activity in the whole of India (of which the English possessions alone amount to one hundred and fifteen million inhabitants) there are not more than three hundred living converts to be found; and at the same time it is admitted that the Christian converts are distinguished for their extreme immorality. There are only three hundred venal and bribed souls out of so many millions. I cannot see that it has gone better with Christianity in India since then, although the missionaries are now trying, contrary to agreement, to work on the children's minds in schools exclusively devoted to secular English instruction, in order to smuggle in Christianity, against which, however, the Hindoos are most jealously on their guard. For, as has been said, childhood is the time, and not manhood, to sow the seeds of belief, especially where an earlier belief has taken root. An acquired conviction, however, that is assumed by matured converts serves, generally, as only the mask for some kind of personal interest. And it is the feeling that this could hardly be otherwise that makes a man, who changes his religion at maturity, despised by most people everywhere; a fact which reveals that they do not regard religion as a matter of reasoned conviction but merely as a belief inoculated in early childhood, before it has been put to any test. That they are right in looking at religion in this way is to be gathered from the fact that it is not only the blind, credulous masses, but also the clergy of every religion, who, as such, have studied its sources, arguments, dogmas and differences, who cling faithfully and zealously as a body to the religion of their fatherland; consequently it is the rarest thing in the world for a priest to change from one religion or creed to another. For instance, we see that the Catholic clergy are absolutely convinced of the truth of all the principles of their Church, and that the Protestants are also of theirs, and that both defend the principles of their confession with like zeal. And yet the conviction is the outcome merely of the country in which each is born: the truth of the Catholic dogma is perfectly clear to the clergy of South Germany, the Protestant to the clergy of North Germany. If, therefore, these convictions rest on objective reasons, these reasons must be climatic and thrive like plants, some only here, some only there. The masses everywhere, however, accept on trust and faith the convictions of those who are *locally convinced*.

Demop. That doesn't matter, for essentially it makes no difference. For instance, Protestantism in reality is more suited to the north, Catholicism to the south.

Phil. So it appears. Still, I take a higher point of view, and have before me a more important object, namely, the progress of the knowledge of truth among the human race. It is a frightful condition of things that, wherever a man is born, certain propositions are inculcated in his earliest youth, and he is assured that under penalty of forfeiting eternal salvation he may never entertain any doubt about them; in so far, that is, as they are propositions which influence the foundation of all our other knowledge and accordingly decide for ever our point of view, and if they are false, upset it for ever. Further, as the influences drawn from these propositions make inroads everywhere into the entire system of our knowledge, the whole of human knowledge is through and through affected by them. This is proved by every literature, and most conspicuously by that of the Middle Age, but also, in too great an extent, by that of the fifteenth and sixteenth centuries. We see how paralysed even the minds of the first rank of all those epochs were by such false fundamental conceptions; and how especially all insight into the true substance and working of Nature was hemmed in on every side. During the whole of the Christian period Theism lay like a kind of oppressive nightmare on all intellectual effort, and on philosophical effort in particular, hindering and arresting all progress. For the men of learning of those epochs, God, devil, angels, demons, hid the whole of Nature; no investigation was carried out to the end, no matter sifted to the bottom; everything that was beyond the most obvious *causal nexus* was immediately attributed to these; so that, as Pomponatius expressed himself at the time, *Certe philosophi nihil verisimile habent ad haec, quare necesse est, ad Deum, ad angelos et daemones recurrere.* It is true that there is a suspicion of irony in what this man says, as his malice in other ways is known, nevertheless he has expressed the general way of thinking of his age. If any one, on the other hand, possessed that rare elasticity of mind which alone enabled him to free himself from the fetters, his writings, and he himself with them, were burnt; as happened to Bruno and Vanini. But how absolutely paralysed the ordinary mind is by that early metaphysical preparation may be seen most strikingly, and from its most ridiculous side, when it undertakes to criticise the doctrines of a foreign belief. One finds the ordinary man, as a rule, merely trying to carefully prove that the dogmas of the foreign belief do not agree with those of his own; he labours to explain that not only do they not say the same, but certainly do not mean the same thing as his. With that he fancies in his simplicity that he has proved the falsity of the doctrines of the alien belief. It really never occurs to him to ask the question which of the two is right; but his own articles of belief are to him as *a priori* certain principles. The Rev. Mr. Morrison has furnished an amusing example of this kind in vol. xx. of the *Asiatic Journal* wherein he criticises the religion and philosophy of the Chinese.

Demop. So that's your higher point of view. But I assure you that there is a higher still. *Primum vivere, deinde philosophari* is of more comprehensive significance than one supposes at first sight. Before everything else, the raw and wicked tendencies of the masses ought to be restrained, in order to protect them from doing anything that is extremely unjust, or committing cruel, violent, and disgraceful deeds. If one waited until they recognised and grasped the truth one would assuredly come too late. And supposing they had already found truth, it would surpass their powers of comprehension. In any case it would be a mere allegorical investiture of truth, a parable, or a myth that would be of any good to them. There must be, as Kant has said, a public standard of right and virtue, nay, this must at all times flutter high. It is all the same in the end what kind of heraldic figures are represented on it, if they only indicate what is meant. Such an allegorical truth is at all times and everywhere, for mankind at large, a beneficial substitute for an eternally unattainable truth, and in general, for a philosophy which it can never grasp; to say nothing of its changing its form daily, and not having as yet attained any kind of general recognition. Therefore practical aims, my good Philalethes, have in every way the advantage of theoretical.

Phil. This closely resembles the ancient advice of Timaeus of Locrus, the Pythagorean: [Greek: tas psychas apeirgomes pseudesi logois, ei ka mae agaetai alathesi].[13] And I almost suspect that it is your wish, according to the fashion of to-day, to remind me—

"Good friend, the time is near
When we may feast off what is good in peace."

And your recommendation means that we should take care in time, so that the waves of the dissatisfied, raging masses may not disturb us at table. But the whole of this point of view is as false as it is nowadays universally liked and praised; this is why I make haste to put in a protest against it. It is *false* that state, justice, and law cannot be maintained without the aid of religion and its articles of belief, and that justice and police regulations need religion as a complement in order to carry out legislative arrangements. It is *false* if it were repeated a hundred times. For the ancients, and especially the Greeks, furnish us with striking *instantia in contrarium* founded on fact. They had absolutely nothing of what we understand by religion. They had no sacred documents, no dogma to be learnt, and its acceptance advanced by every one, and its principles inculcated early in youth. The servants of religion preached just as little about morals, and the ministers concerned themselves very little about any kind of morality or in general about what the people either did or left undone. No such thing. But the duty of the priests was confined merely to temple ceremonies, prayers, songs, sacrifices, processions, lustrations, and the like, all of which aimed at anything but the moral improvement of the individual. The whole of their so-called religion consisted, and particularly in the towns, in some of the *deorum majorum gentium* having temples here and there, in which the aforesaid worship was conducted as an affair of state, when in reality it was an affair of police. No one, except the functionaries engaged, was obliged in any way to be present, or even to believe in it. In the whole of antiquity there is no trace of any obligation to believe in any kind of dogma. It was merely any one who openly denied the existence of the gods or calumniated them that was punished; because by so doing he insulted the state which served these gods; beyond this every one was allowed to think what he chose of them. If any one wished to win the favour of these gods privately by prayer or sacrifice he was free to do so at his own cost and risk; if he did not do it, no one had anything to say against it, and least of all the State. Every Roman had his own Lares and Penates at home, which were, however, at bottom nothing more than the revered portraits of his ancestors. The ancients had no kind of decisive, clear, and least of all dogmatically fixed ideas about the immortality of the soul and a life hereafter, but every one in his own way had lax, vacillating, and problematical ideas; and their ideas about the gods were just as various, individual, and vague. So that the ancients had really no *religion* in our sense of the word. Was it for this reason that anarchy and lawlessness reigned among them? Is not law and civil order rather so much their work, that it still constitutes the foundation of ours? Was not property perfectly secure, although it consisted of slaves for the greater part? And did not this condition of things last longer than a thousand years?

So I cannot perceive, and must protest against the practical aims and necessity of religion in the sense which you have indicated, and in such general favour to-day, namely, as an indispensable foundation of all legislative regulations. For from such a standpoint the pure and sacred striving after light and truth, to say the least, would seem quixotic and criminal if it should venture in its feeling of justice to denounce the authoritative belief

119

as a usurper who has taken possession of the throne of truth and maintained it by continuing the deception.

Demop. But religion is not opposed to truth; for it itself teaches truth. Only it must not allow truth to appear in its naked form, because its sphere of activity is not a narrow auditory, but the world and humanity at large, and therefore it must conform to the requirements and comprehension of so great and mixed a public; or, to use a medical simile, it must not present it pure, but must as a medium make use of a mythical vehicle. Truth may also be compared in this respect to certain chemical stuffs which in themselves are gaseous, but which for official uses, as also for preservation or transmission, must be bound to a firm, palpable base, because they would otherwise volatilise. For example, chlorine is for all such purposes applied only in the form of chlorides. But if truth, pure, abstract, and free from anything of a mythical nature, is always to remain unattainable by us all, philosophers included, it might be compared to fluorine, which cannot be presented by itself alone, but only when combined with other stuffs. Or, to take a simpler simile, truth, which cannot be expressed in any other way than by myth and allegory, is like water that cannot be transported without a vessel; but philosophers, who insist upon possessing it pure, are like a person who breaks the vessel in order to get the water by itself. This is perhaps a true analogy. At any rate, religion is truth allegorically and mythically expressed, and thereby made possible and digestible to mankind at large. For mankind could by no means digest it pure and unadulterated, just as we cannot live in pure oxygen but require an addition of four-fifths of nitrogen. And without speaking figuratively, the profound significance and high aim of life can only be revealed and shown to the masses symbolically, because they are not capable of grasping life in its real sense; while philosophy should be like the Eleusinian mysteries, for the few, the elect.

Phil. I understand. The matter resolves itself into truth putting on the dress of falsehood. But in doing so it enters into a fatal alliance. What a dangerous weapon is given into the hands of those who have the authority to make use of falsehood as the vehicle of truth! If such is the case, I fear there will be more harm caused by the falsehood than good derived from the truth. If the allegory were admitted to be such, I should say nothing against it; but in that case it would be deprived of all respect, and consequently of all efficacy. Therefore the allegory must assert a claim, which it must maintain, to be true in *sensu proprio* while at the most it is true in *sensu allegorico*. Here lies the incurable mischief, the permanent evil; and therefore religion is always in conflict, and always will be with the free and noble striving after pure truth.

Demop. Indeed, no. Care has been taken to prevent that. If religion may not exactly admit its allegorical nature, it indicates it at any rate sufficiently.

Phil. And in what way does it do that?

Demop. In its mysteries. *Mystery* is at bottom only the theological *terminus technicus* for religious allegory. All religions have their mysteries. In reality, a mystery is a palpably absurd dogma which conceals in itself a lofty truth, which by itself would be absolutely incomprehensible to the ordinary intelligence of the raw masses. The masses accept it in this disguise on trust and faith, without allowing themselves to be led astray by its absurdity, which is palpable to them; and thereby they participate in the kernel of the matter so far as they are able. I may add as an explanation that the use of mystery has been attempted even in philosophy; for example, when Pascal, who was pietest, mathematician, and philosopher in one, says in this threefold character: *God is everywhere centre and nowhere periphery.* Malebranche has also truly remarked, *La liberté est un mystère.* One might go further, and maintain that in religions everything is really mystery. For it is utterly impossible to impart truth in *sensu proprio* to the multitude in its crudity; it is only a mythical and allegorical reflection of it that can fall to its share and enlighten it. Naked truth must not appear before the eyes of the profane vulgar; it can only appear before them closely veiled. And it is for this reason that it is unfair to demand of a religion that it should be true in *sensu proprio*, and that, *en passant.* Rationalists and Supernaturalists of to-day are so absurd. They both start with the supposition that religion must be the truth; and while the former prove that it is not, the latter obstinately maintain that it is; or rather the former cut up and dress the allegory in such a way that it could be true in *sensu proprio* but would in that case become a platitude. The latter wish to maintain, without further dressing, that it is true in *sensu proprio*, which, as they should know, can only be carried into execution by inquisitions and the stake. While in reality, myth and allegory are the essential elements of religion, but under the indispensable condition (because of the intellectual limitations of the great masses) that it supplies enough satisfaction to meet those metaphysical needs of mankind which are ineradicable, and that it takes the place of pure philosophical truth, which is infinitely difficult, and perhaps never attainable.

Phil. Yes, pretty much in the same way as a wooden leg takes the place of a natural one. It supplies what is wanting, does very poor service for it, and claims to be regarded as a natural leg, and is more or less cleverly put together. There is a difference, however, for, as a rule, the natural leg was in existence before the wooden one, while religion everywhere has gained the start of philosophy.

Demop. That may be; but a wooden leg is of great value to those who have no natural leg. You must keep in view that the metaphysical requirements of man absolutely demand satisfaction; because the horizon of his thoughts must be defined and not remain unlimited. A man, as a rule, has no faculty of judgment for weighing reasons, and distinguishing between what is true and what is false. Moreover, the work imposed upon him by nature and her requirements leaves him no time for investigations of that kind, or for the education which they presuppose. Therefore it is entirely out of the question to imagine he will be convinced by reasons; there is nothing left for him but belief and authority. Even if a really true philosophy took the place of religion, at least nine-tenths of mankind would only accept it on authority, so that it would be again a matter of belief; for Plato's [Greek: philosophon plaethos adynaton einai] will always hold good. Authority, however, is only established by time and circumstances, so that we cannot bestow it on that which has only reason to commend it; accordingly, we must grant it only to that which has attained it in the course of history, even if it is only truth represented allegorically. This kind of truth, supported by authority, appeals directly to the essentially metaphysical temperament of man—that is, to his need of a theory concerning the riddle of existence, which thrusts itself upon him, and arises from the consciousness that behind the physical in the world there must be a metaphysical, an unchangeable something, which serves as the foundation of constant change. It also appeals to the will, fears, and hopes of mortals living in constant need; religion provides them with gods, demons, to whom they call, appease, and conciliate. Finally, it appeals to their moral consciousness, which is undeniably present, and lends to it that authenticity and support from without—a support without which it would not easily maintain itself in the struggle against so many temptations. It is exactly from this side that religion provides an inexhaustible source of consolation and comfort in the countless and great sorrows of life, a comfort which does not leave men in death, but rather then unfolds its full efficacy. So that religion is like some one taking hold of the hand of a blind person and leading him, since he cannot see for himself; all that the blind person wants is to attain his end, not to see everything as he walks along.

Phil. This side is certainly the brilliant side of religion. If it is a *fraus* it is indeed a *pia fraus*; that cannot be denied. Then priests become something between deceivers and moralists. For they dare not teach the real truth, as you yourself have quite correctly explained, even if it were known to them; which it is not. There can, at any rate, be a true philosophy, but there can be no true religion: I mean true in the real and proper understanding of the word, not merely in that flowery and allegorical sense which you have described, a sense in which every religion would be true only in different degrees. It is certainly quite in harmony with the inextricable admixture of good and evil, honesty and dishonesty, goodness and wickedness, magnanimity and baseness, which the world presents everywhere, that the most important, the most lofty, and the most sacred truths can make their appearance only in combination with a lie, nay, can borrow strength from a lie as something that affects mankind more powerfully; and as revelation must be introduced by a lie. One might regard this fact as the *monogram* of the moral world. Meanwhile let us not give up the hope that mankind will some day attain that point of maturity and education at which it is able to produce a true philosophy on the one hand, and accept it on the other. *Simplex sigillum veri*: the naked truth must be so simple and comprehensible that one can impart it to all in its true form without any admixture of myth and fable (a pack of lies)—in other words, without masking it as *religion*.

Demop. You have not a sufficient idea of the wretched capacities of the masses.

Phil. I express it only as a hope; but to give it up is impossible. In that case, if truth were in a simpler and more comprehensible form, it would surely soon drive religion from the position of vicegerent which it has so long held. Then religion will have fulfilled her mission and finished her course; she might then dismiss the race which she has guided to maturity and herself retire in peace. This will be the *euthanasia* of religion. However, as long as she lives she has two faces, one of truth and one of deceit. According as one looks attentively at one or the other one will like or dislike her. Hence religion must be regarded as a necessary evil, its necessity resting on the pitiful weak-mindedness of the great majority of mankind, incapable of grasping the truth, and consequently when in extremity requires a substitute for truth.

Demop. Really, one would think that you philosophers had truth lying in readiness, and all that one had to do was to lay hold of it.

Phil. If we have not got it, it is principally to be ascribed to the pressure under which philosophy, at all periods and in all countries, has been held by religion. We have tried to make not only the expression and communication of truth impossible, but even the contemplation and discovery of it, by giving the minds of children in earliest childhood into the hands of priests to be worked upon; to have the groove in which their fundamental thoughts are henceforth to run so firmly imprinted, as in principal matters, to become fixed and determined for a lifetime. I am sometimes shocked to see when I take into my hand the writings of even the most intelligent minds of the sixteenth and seventeenth centuries, and especially if I have just left my oriental studies, how paralysed and hemmed in on all sides they are by Jewish notions. Prepared in this way, one cannot form any idea of the true philosophy!

Demop. And if, moreover, this true philosophy were discovered, religion would not cease to exist, as you imagine. There cannot be one system of metaphysics for everybody; the natural differences of intellectual power in addition to those of education make this impossible. The great majority of mankind must necessarily be engaged in that arduous bodily labour which is requisite in order to furnish the endless needs of the whole race. Not only does this leave the majority no time for education, for learning, or for reflection; but by virtue of the strong antagonism between merely physical and intellectual qualities, much excessive bodily labour blunts the understanding and makes it heavy, clumsy, and awkward, and consequently incapable of grasping any other than perfectly simple and palpable matters. At least nine-tenths of the human race comes under this category. People require a system of metaphysics, that is, an account of the world and our existence, because such an account belongs to the most natural requirements of mankind. They require also a popular system of metaphysics, which, in order for it to be this, must combine many rare qualities; for instance, it must be exceedingly lucid, and yet in the right places be obscure, nay, to a certain extent, impenetrable; then a correct and satisfying moral system must be combined with its dogmas; above everything, it must bring inexhaustible consolation in suffering and death. It follows from this that it can only be true in *sensu allegorico* and not in *sensu proprio*. Further, it must have the support of an authority which is imposing by its great age, by its general recognition, by its documents, together with their tone and statements—qualities which are so infinitely difficult to combine that many a man, if he stopped to reflect, would not be so ready to help to undermine a religion, but would consider it the most sacred treasure of the people. If any one wants to criticise religion he should always bear in mind the nature of the great masses for which it is destined, and picture to himself their complete moral and intellectual inferiority. It is incredible how far this inferiority goes and how steadily a spark of truth will continue to glimmer even under the crudest veiling of monstrous fables and grotesque ceremonies, adhering indelibly, like the perfume of musk, to everything which has come in contact with it. As an illustration of this, look at the profound wisdom which is revealed in the Upanishads, and then look at the mad idolatry in the India of to-day, as is revealed in its pilgrimages, processions, and festivities, or at the mad and ludicrous doings of the Saniassi of the present time. Nevertheless, it cannot be denied that in all this madness and absurdity there yet lies something that is hidden from view, something that is in accordance with, or a reflection of the profound wisdom that has been mentioned. It requires this kind of dressing-up for the great brute masses. In this antithesis we have before us the two poles of humanity:—the wisdom of the individual and the bestiality of the masses, both of which, however, find their point of harmony in the moral kingdom. Who has not thought of the saying from the Kurral—"Vulgar people look like men; but I have never seen anything like them." The more highly cultured man may always explain religion to himself *cum grano salis*; the man of learning, the thoughtful mind, may, in secret, exchange it for a philosophy. And yet *one* philosophy would not do for everybody; each philosophy by the laws of affinity attracts a public to whose education and mental capacities it is fitted. So there is always an inferior metaphysical system of the schools for the educated plebeians, and a higher system for the *elite*. Kant's lofty doctrine, for example, was degraded to meet the requirements of the schools, and ruined by Fries, Krug, Salat, and similar people. In short, Goethe's dictum is as applicable here as anywhere: *One does not suit all.* Pure belief in revelation and pure metaphysics are for the two extremes; and for the intermediate steps mutual modifications of both in countless combinations and gradations. The immeasurable differences which nature and education place between men have made this necessary.

Phil. This point of view reminds me seriously of the mysteries of the ancients which you have already mentioned; their aim at bottom seems to have lain in remedying the evil arising out of the differences of mental capacities and education. Their plan was to single out of the great multitude a few people, to whom the unveiled truth was absolutely incomprehensible, and to reveal the truth to them up to a certain point; then out of these they singled out others to whom they revealed more, as they were able to grasp more; and so on up to the Epopts. And so we got [Greek: mikra, kai meizona, kai megista mystaeria]. The plan was based on a correct knowledge of the intellectual inequality of mankind.

Demop. To a certain extent the education in our lower, middle, and high schools represents the different forms of initiation into the mysteries.

Phil. Only in a very approximate way, and this only in so far as subjects of higher knowledge were written about exclusively in Latin. But since that has ceased to be so all the mysteries are profaned.

Demop. However that may be, I wish to remind you, in speaking of religion, that you should grasp it more from the practical and less from the theoretical side. Personified metaphysics may be religion's enemy, yet personified morality will be its friend. Perhaps the metaphysics in all religions is false; but the morality in all is true. This is to be surmised from the fact that in their metaphysics they contradict each other, while in their morality they agree.

Phil. Which furnishes us with a proof of the rule of logic, that a true conclusion may follow from false premises.

Demop. Well, stick to your conclusion, and be always mindful that religion has two sides. If it can't stand when looked at merely from the theoretical—in other words, from its intellectual side, it appears, on the other hand, from the moral side as the only means of directing, training, and pacifying those races of animals gifted with reason, whose kinship with the ape does not exclude a kinship with the tiger. At the same time religion is, in general, a sufficient satisfaction for their dull metaphysical needs. You appear to me to have no proper idea of the difference, wide as the heavens apart, of the profound breach between your learned man, who is enlightened and accustomed to think, and the heavy, awkward, stupid, and inert consciousness of mankind's beasts of burden, whose thoughts have taken once and for all the direction of fear about their maintenance, and cannot be put in motion in any other; and whose muscular power is so exclusively exercised that the nervous power which produces intelligence is thereby greatly reduced. People of this kind must absolutely have something that they can take hold of on the slippery and thorny path of their life, some sort of beautiful fable by means of which things can be presented to them which their crude intelligence could most certainly only understand in picture and parable. It is impossible to approach them with subtle explanations and fine distinctions. If you think of religion in this way, and bear in mind that its aims are extremely practical and only subordinately theoretical, it will seem to you worthy of the highest respect.

Phil. A respect which would finally rest on the principle that the end sanctifies the means. However, I am not in favour of a compromise on a basis of that sort. Religion may be an excellent means of curbing and controlling the perverse, dull, and malicious creatures of the biped race; in the eyes of the friend of truth every *fraus*, be it ever so *pia*, must be rejected. It would be an odd way to promote virtue through the medium of lies and deception. The flag to which I have sworn is truth. I shall remain faithful to it everywhere, and regardless of success, I shall fight for light and truth. If I see religion hostile, I shall—

Demop. But you will not! Religion is not a deception; it is true, and the most important of all truths. But because, as has already been said, its doctrines are of such a lofty nature that the great masses cannot grasp them immediately; because, I say, its light would blind the ordinary eye, does it appear concealed in the veil of allegory and teach that which is not exactly true in itself, but which is true according to the meaning contained in it: and understood in this way religion is the truth.

Phil. That would be very probable, if it were allowed to be true only in an allegorical sense. But it claims to be exactly true, and true in the proper sense of the word: herein lies the deception, and it is here that the friend of truth must oppose it.

Demop. But this deception is a *conditio sine qua non.* If religion admitted that it was merely the allegorical meaning in its doctrines that was true, it would be deprived of all efficacy, and such rigorous treatment would put an end to its invaluable and beneficial influence on the morals and feelings of mankind. Instead of insisting on that with pedantic obstinacy, look at its great achievements in a practical way both as regards morality and feelings, as a guide to conduct, as a support and consolation to suffering humanity in life and death. How greatly you should guard against rousing suspicion in the masses by theoretical wrangling, and thereby finally taking from them what is an inexhaustible source of consolation and comfort to them; which in their hard lot they need very much more than we do: for this reason alone, religion ought not to be attacked.

Phil. With this argument Luther could have been beaten out of the field when he attacked the selling of indulgences; for the letters of indulgence have furnished many a man with irreparable consolation and perfect tranquillity, so that he joyfully passed away with perfect confidence in the little packet of them which he firmly held in his hand as he lay dying, convinced that in them he had so many cards of admission into all the nine heavens. What is the use of grounds of consolation and peacefulness over which is constantly hanging the Damocles-sword of deception? The truth, my friend, the truth alone holds good, and remains constant and faithful; it is the only solid consolation; it is the indestructible diamond.

Demop. Yes, if you had truth in your pocket to bless us with whenever we asked for it. But what you possess are only metaphysical systems in which nothing is certain but the headaches they cost. Before one takes anything away one must have something better to put in its place.

Phil. I wish you would not continually say that. To free a man from error does not mean to take something from him, but to give him something. For knowledge that something is wrong is a truth. No error, however, is harmless; every error will cause mischief sooner or later to the man who fosters it. Therefore do not deceive any one, but rather admit you are ignorant of what you do not know, and let each man form his own dogmas for himself. Perhaps they will not turn out so bad, especially as they will rub against each other and mutually rectify errors; at any rate the various opinions will establish tolerance. Those men who possess both knowledge and capacity may take up the study of philosophy, or even themselves advance the history of philosophy.

Demop. That would be a fine thing! A whole nation of naturalised metaphysicians quarrelling with each other, and *eventualiter* striking each other.

Phil. Well, a few blows here and there are the sauce of life, or at least a very slight evil compared with priestly government—prosecution of heretics, plundering of the laity, courts of inquisition, crusades, religious wars, massacres of St. Bartholomew, and the like. They have been the results of chartered popular metaphysics: therefore I still hold that one cannot expect to get grapes from thistles, or good from lies and deception.

Demop. How often must I repeat that religion is not a lie, but the truth itself in a mythical, allegorical dress? But with respect to your plan of each man establishing his own religion, I had still something to say to you, that a particularism like this is totally and absolutely opposed to the nature of mankind, and therefore would abolish all social order. Man is an *animal metaphysicum*—in other words, he has surpassingly great metaphysical requirements; accordingly he conceives life above all in its metaphysical sense, and from that standpoint wishes to grasp everything. Accordingly, odd as it may sound with regard to the uncertainty of all dogmas, accord in the fundamental elements of metaphysics is the principal thing, in so much as it is only among people who hold the same views on this question that a genuine and lasting fellowship is possible. As a result of this, nations resemble and differ from each other more in religion than in government, or even language. Consequently, the fabric of society, the State, will only be perfectly firm when it has for a basis a system of metaphysics universally acknowledged. Such a system, naturally, can only be a popular metaphysical one—that is, a religion. It then becomes identified with the government, with all the general expressions of the national life, as well as with all sacred acts of private life. This was the case in ancient India, among the Persians, Egyptians, Jews, also the Greeks and Romans, and it is still the case among the Brahman, Buddhist, and Mohammedan nations. There, are three doctrines of faith in China, it is true, and the one that has spread the most, namely, Buddhism, is exactly the doctrine that is least protected by the State; yet there is a saying in China that is universally appreciated and daily applied, *the three doctrines are only one*—in other words, they agree in the main thing. The Emperor confesses all three at the same time, and agrees with them all. Europe is the confederacy of *Christian* States; Christianity is the basis of each of its members and the common bond of all; hence Turkey, although it is in Europe, is really not to be reckoned in it. Similarly the European princes are such "by the grace of God," and the Pope is the delegate of God; accordingly, as his throne was the highest, he wished all other thrones to be looked upon only as held in fee from him. Similarly Archbishops and Bishops, as such, had temporal authority, just as they have still in England a seat and voice in the Upper House; Protestant rulers are, as such, heads of their churches; in England a few years ago this was a girl of eighteen. By the revolt from the Pope, the Reformation shattered the European structure, and, in particular, dissolved the true unity of Germany by abolishing its common faith; this unity, which had as a matter of fact come to grief, had accordingly to be replaced later by artificial and purely political bonds. So you see how essentially connected is unity of faith with common order and every state. It is everywhere the support of the laws and the constitution—that is to say, the foundation of the social structure, which would stand with difficulty if faith did not lend power to the authority of the government and the importance of the ruler.

Phil. Oh, yes, princes look upon God as a goblin, wherewith to frighten grown-up children to bed when nothing else is of any avail; it is for this reason that they depend so much on God. All right; meanwhile I should like to advise every ruling lord to read through, on a certain day every six months, the fifteenth chapter of the First Book of Samuel, earnestly and attentively; so that he may always have in mind what it means to support the throne on the altar. Moreover, since burning at the stake, that *ultima ratio theologorum*, is a thing of the past, this mode of government has lost its efficacy. For, as you know, religions are like glowworms: before they can shine it must be dark. A certain degree of general ignorance is the condition of every religion, and is the element in which alone it is able to exist. While, as soon as astronomy, natural science, geology, history, knowledge of countries and nations have spread their light universally, and philosophy is finally allowed to speak, every faith which is based on miracle and revelation must perish, and then philosophy will take its place. In Europe the day of knowledge and science dawned towards the end of the fifteenth century with the arrival of the modern Greek philosophers, its sun rose higher in the sixteenth and seventeenth centuries, which were so productive, and scattered the mists of the Middle Age. In the same proportion, both Church and Faith were obliged to gradually disappear; so that in the eighteenth century English and French philosophers became direct antagonists, until finally, under Frederick the Great, Kant came and took away from religious belief the support it had formerly received from philosophy, and emancipated the *ancilla theologiae* in that he attacked the question with German thoroughness and perseverance, whereby it received a less frivolous, that is to say, a more earnest tone. As a result of this we see in the nineteenth century Christianity very much weakened, almost stripped entirely of serious belief, nay, fighting for its own existence; while apprehensive princes try to raise it up by an artificial stimulant, as the doctor tries to revive a dying man by the aid of a drug. There is a passage from Condorcet's *Des Progres de l'esprit humain*, which seems to have been written as a warning to our epoch: *Le zele religieux des philosophes et des grands n'etait qu'une devotion politique: et toute religion, qu'on se permet de defendre comme une croyance qu'il est utile de laisser au peuple, ne peut plus esperer qu'une agonie plus ou moins prolongee.* In the whole course of the events which I have pointed out you may always observe that belief and knowledge bear the same relation to each other as the two scales of a balance: when the one rises the other must fall. The balance is so sensitive that it indicates momentary influences. For example, in the beginning of this century the predatory excursions of French robbers under their leader Buonaparte, and the great efforts that were requisite to drive them out and to punish them, had led to a temporary neglect of science, and in consequence to a certain decrease in the general propagation of knowledge; the Church immediately began to raise her head again and Faith to be revived, a revival partly of a poetical nature, in keeping with the spirit of the times. On the other hand, in the more than thirty years' peace that followed, leisure and prosperity promoted the building up of science and the spread of knowledge in an exceptional degree, so that the result was what I have said, the dissolution and threatened fall of religion. Perhaps the time which has been so often predicted is not far distant, when religion will depart from European humanity, like a nurse whose care the child has outgrown; it is now placed in the hands of a tutor for instruction. For without doubt doctrines of belief that are based only on authority, miracles, and revelation are only of use and suitable to the childhood of humanity. That a race, which all physical and historical data confirm as having been in existence only about a hundred times the life of a man sixty years old, is still in its first childhood is a fact that every one will admit.

Demop. If instead of prophesying with undisguised pleasure the downfall of Christianity, you would only consider how infinitely indebted European humanity is to it, and to the religion which, after the lapse of some time, followed Christianity from its old home in the East! Europe received from it a drift which had hitherto been unknown to it—it learnt the fundamental truth that life cannot be an end-in-itself, but that the true end of our existence lies beyond it. The Greeks and Romans had placed this end absolutely in life itself, so that, in this sense, they may most certainly be called blind heathens. Correspondingly, all their virtues consist in what is serviceable to the public, in what is useful; and Aristotle says quite naively, *"Those virtues must necessarily be the greatest which are the most useful to others"* ([Greek: anankae de megistas einai aretas tas tois allois chraesimotatas], *Rhetor.* I. c. 9). This is why the ancients considered love for one's country the greatest virtue, although it is a very doubtful one, as it is made up of narrowness, prejudice, vanity, and an enlightened self-interest. Preceding the passage that has just been quoted, Aristotle enumerates all the virtues in order to explain them individually. They are *Justice, Courage, Moderation, Magnificence* ([Greek: megaloprepeia]), *Magnanimity, Liberality, Gentleness, Reasonableness, and Wisdom.* How different from the Christian virtues! Even Plato, without comparison the most transcendental philosopher of pre-Christian antiquity, knows no higher virtue than *Justice*; he alone recommends it unconditionally and for its own sake, while all the other philosophers make a happy life—*vita beata*—the aim of all virtue; and it is acquired through the medium of moral behaviour. Christianity released European humanity from its superficial and crude absorption in an ephemeral, uncertain, and hollow existence.

> . . . *coelumque tueri*
> *Jussit, et erectos ad sidera tollere vultus.*

Accordingly, Christianity does not only preach Justice, but the *Love of Mankind, Compassion, Charity, Reconciliation, Love of one's Enemies, Patience, Humility, Renunciation, Faith, and Hope*. Indeed, it went even further: it taught that the world was of evil and that we needed deliverance; consequently it preached contempt of the world, self-denial, chastity, the giving up of one's own will, that is to say, turning away from life and its phantom-like pleasures; it taught further the healing power of suffering, and that an instrument of torture is the symbol of Christianity, I willingly admit that this serious and only correct view of life had spread in other forms throughout Asia thousands of years previously, independently of Christianity as it is still; but this view of life was a new and tremendous revelation to European humanity. For it is well known that the population of Europe consists of Asiatic races who, driven out from their own country, wandered away, and by degrees hit upon Europe: on their long wanderings they lost the original religion of their homes, and with it the correct view of life; and this is why they formed in another climate religions for themselves which were somewhat crude; especially the worship of Odin, the Druidic and the Greek religions, the metaphysical contents of which were small and shallow. Meanwhile there developed among the Greeks a quite special, one might say an instinctive, sense of beauty, possessed by them alone of all the nations of the earth that have ever existed—a peculiar, fine, and correct sense of beauty, so that in the mouths of their poets and in the hands of their artists, their mythology took an exceptionally beautiful and delightful form. On the other hand, the earnest, true, and profound import of life was lost to the Greeks and Romans; they lived like big children until Christianity came and brought them back to the serious side of life.

Phil. And to form an idea of the result we need only compare antiquity with the Middle Age that followed—that is, the time of Pericles with the fourteenth century. It is difficult to believe that we have the same kind of beings before us. There, the finest development of humanity, excellent constitutional regulations, wise laws, cleverly distributed offices, rationally ordered freedom, all the arts, as well as poetry and philosophy, at their best; the creation of works which after thousands of years have never been equalled and are almost works of a higher order of beings, whom we can never approach; life embellished by the noblest fellowship, as is portrayed in the *Banquet* of Xenophon. And now look at this side, if you can. Look at the time when the Church had imprisoned the minds, and violence the bodies of men, whereby knights and priests could lay the whole weight of life on the common beast of burden—the third estate. There you have club-law, feudalism, and fanaticism in close alliance, and in their train shocking uncertainty and darkness of mind, a corresponding intolerance, discord of faiths, religious wars, crusades, persecution of heretics and inquisitions; as the form of fellowship, chivalry, an amalgam of savagery and foolishness, with its pedantic system of absurd affectations, its degrading superstitions, and apish veneration for women; the survival of which is gallantry, deservedly requited by the arrogance of women; it affords to all Asiatics continual material for laughter, in which the Greeks would have joined. In the golden Middle Age the matter went as far as a formal and methodical service of women and enjoined deeds of heroism, *cours d'amour*, bombastic Troubadour songs and so forth, although it is to be observed that these last absurdities, which have an intellectual side, were principally at home in France; while among the material phlegmatic Germans the knights distinguished themselves more by drinking and robbing. Drinking and hoarding their castles with plunder were the occupations of their lives; and certainly there was no want of stupid love-songs in the courts. What has changed the scene so? Migration and Christianity.

Demop. It is a good thing you reminded me of it. Migration was the source of the evil, and Christianity the dam on which it broke. Christianity was the means of controlling and taming those raw, wild hordes who were washed in by the flood of migration. The savage man must first of all learn to kneel, to venerate, and to obey; it is only after that, that he can be civilised. This was done in Ireland by St. Patrick, in Germany by Winifred the Saxon, who was a genuine Boniface. It was migration of nations, this last movement of Asiatic races towards Europe, followed only by their fruitless attempts under Attila, Gengis Khan, and Timur, and, as a comic after-piece, by the gipsies: it was migration of nations which swept away the humanity of the ancients. Christianity was the very principle which worked against this savagery, just as later, through the whole of the Middle Age, the Church and its hierarchy were extremely necessary to place a limit to the savagery and barbarism of those lords of violence, the princes and knights: it was the ice-breaker of this mighty flood. Still, the general aim of Christianity is not so much to make this life pleasant as to make us worthy of a better. It looks beyond this span of time, this fleeting dream, in order to lead us to eternal salvation. Its tendency is ethical in the highest sense of the word, a tendency which had hitherto been unknown in Europe; as I have already pointed out to you by comparing the morality and religion of the ancients with those of Christianity.

Phil. That is right so far as theory is concerned; but look at the practice. In comparison with the Christian centuries that followed, the ancient world was undoubtedly less cruel than the Middle Age, with its deaths by frightful torture, its countless burnings at the stake; further, the ancients were very patient, thought very highly of justice, and frequently sacrificed themselves for their country, showed traits of magnanimity of every kind, and such genuine humanity, that, up to the present time, an acquaintance with their doings and thoughts is called the study of Humanity. Religious wars, massacres, crusades, inquisitions, as well as other persecutions, the extermination of the original inhabitants of America and the introduction of African slaves in their place, were the fruits of Christianity, and among the ancients one cannot find anything analogous to this, anything to counterpoise it; for the slaves of the ancients, the *familia*, the *vernae*, were a satisfied race and faithfully devoted to their masters, and as widely distinct from the miserable negroes of the sugar plantations, which are a disgrace to humanity, as they were in colour. The censurable toleration of pederasty, for which one chiefly reproaches the morality of the ancients, is a trifle compared with the Christian horrors I have cited, and is not so rare among people of to-day as it appears to be. Can you then, taking everything into consideration, maintain that humanity has really become morally better by Christianity?

Demop. If the result has not everywhere corresponded with the purity and accuracy of the doctrine, it may be because this doctrine has been too noble, too sublime for humanity, and its aim set too high: to be sure, it was much easier to comply with heathen morality or with the Mohammedan. It is precisely what is most elevated that is the most open to abuse and deception—*abusus optimi pessimus*; and therefore those lofty doctrines have sometimes served as a pretext for the most disgraceful transactions and veritable crimes. The downfall of the ancient institutions, as well as of the arts and sciences of the old world, is, as has been said, to be ascribed to the invasion of foreign barbarians. Accordingly, it was inevitable that ignorance and savagery got the upper hand; with the result that violence and fraud usurped their dominion, and knights and priests became a burden to mankind. This is partly to be explained by the fact that the new religion taught the lesson of eternal and not temporal welfare, that simplicity of heart was preferable to intellectual knowledge, and it was averse to all worldly pleasures which are served by the arts and sciences. However, in so far as they could be made serviceable to religion they were promoted, and so flourished to a certain extent.

Phil. In a very narrow sphere. The sciences were suspicious companions, and as such were placed under restrictions; while fond ignorance, that element so necessary to the doctrines of faith, was carefully nourished.

Demop. And yet what humanity had hitherto acquired in the shape of knowledge, and handed down in the works of the ancients, was saved from ruin by the clergy, especially by those in the monasteries. What would have happened if Christianity had not come in just before the migration of nations?

Phil. It would really be an extremely useful inquiry if some one, with the greatest frankness and impartiality, tried to weigh exactly and accurately the advantages and disadvantages derived from religions. To do this, it would be necessary to have a much greater amount of historical and psychological data than either of us has at our command. Academies might make it a subject for a prize essay.

Demop. They will take care not to do that.

Phil. I am surprised to hear you say that, for it is a bad look-out for religion. Besides, there are also academies which make it a secret condition in submitting their questions that the prize should be given to the competitor who best understands the art of flattering them. If we, then, could only get a statistician to tell us how many crimes are prevented yearly by religious motives, and how many by other motives. There would be very few of the former. If a man feels himself tempted to commit a crime, certainly the first thing which presents itself to his mind is the punishment he must suffer for it, and the probability that he will be punished; after that comes the second consideration, that his reputation is at stake. If I am not mistaken, he will reflect by the hour on these two obstacles before religious considerations ever come into his mind. If he can get away from these two first safeguards against crime, I am convinced that religion *alone* will very rarely keep him back from it.

Demop. I believe, however, that it will do so very often; especially when its influence works through the medium of custom, and thereby immediately makes a man shrink from the idea of committing a crime. Early impressions cling to him. As an illustration of what I mean, consider how many a man, and especially if he is of noble birth, will often, in order to fulfil some promise, make great sacrifices, which are instigated solely by the fact that his father has often impressed it upon him in childhood that "a man of honour, or a gentleman, or a cavalier, always keeps his word inviolate."

Phil. And that won't work unless there is a certain innate *probitas*. You must not ascribe to religion what is the result of innate goodness of character, by which pity for the one who would be affected by the crime prevents a man from committing it. This is the genuine moral motive, and as such it is independent of all religions.

Demop. But even this moral motive has no effect on the masses unless it is invested with a religious motive, which, at any rate, strengthens it. However, without any such natural foundation, religious motives often in themselves alone prevent crime: this is not a matter of surprise to us in the case of the multitude, when we see that even people of good education sometimes come under the influence, not indeed of religious motives, which fundamentally are at least allegorically true, but of the most absurd superstitions, by which they are guided throughout the whole of their lives; as, for instance, undertaking nothing on a Friday, refusing to sit down thirteen at table, obeying chance omens, and the like: how much more likely are the masses to be guided by such things. You cannot properly conceive the great limitations of the raw mind; its interior is entirely dark, especially if, as is often the case, a bad, unjust, and wicked heart is its foundation. Men like these, who represent the bulk of humanity, must be directed and controlled meanwhile, as well as possible, even if it be by really superstitious motives, until they become susceptible to truer and better ones. Of the direct effect of religion, one may give as an instance a common occurrence in Italy, namely, that of a thief being allowed to replace what he has stolen through the medium of his confessor, who makes this the condition of his absolution. Then think of the case of an oath, where religion shows a most decided influence: whether it be because a man places himself expressly in the position of a mere *moral being*, and as such regards himself as solemnly appealed to,—as seems to be the case in France, where the form of the oath is merely "*je le jure*"; and among the Quakers, whose solemn "yea" or "nay" takes the place of the oath;—or whether it is because a man really believes he is uttering something that will forfeit his eternal happiness,—a belief which is obviously only the investiture of the former feeling. At any rate, religious motives are a means of awakening and calling forth his moral nature. A man will frequently consent to take a false oath, but suddenly refuse to do so when it comes to the point; whereby truth and right come off victorious.

Phil. But false oaths are still oftener sworn, whereby truth and right are trodden underfoot with the clear knowledge of all the witnesses of the act. An oath is the jurist's metaphysical *pons asinorum*, and like this should be used as seldom as ever possible. When it cannot be avoided, it should be taken with great solemnity, always in the presence of the clergy—nay, even in a church or in a chapel adjoining the court of justice . . . This is precisely why the French abstract formulary of the oath is of no value. By the way, you are right to cite the oath as an undeniable example of the practical efficacy of religion. I must, in spite of everything you have said, doubt whether the efficacy of religion goes much beyond this. Just think, if it were suddenly declared by public proclamation that all criminal laws were abolished; I believe that neither you nor I would have the courage to go home from here alone under the protection of religious motives. On the other hand, if in a similar way all religions were declared to be untrue; we would, under the protection of the laws alone, live on as formerly, without any special increase in our fears and measures of precaution. But I will even go further: religions have very frequently a decidedly demoralising influence. It may be said generally that duties towards God are the reverse of duties towards mankind; and that it is very easy to make up for lack of good behaviour towards men by adulation of God. Accordingly, we see in all ages and countries that the great majority of mankind find it much easier to beg admission into Heaven by prayers than to deserve it by their actions. In every religion it soon comes to be proclaimed that it is not so much moral actions as faith, ceremonies, and rites of every kind that are the immediate objects of the Divine will; and indeed the latter, especially if they are bound up with the emoluments of the clergy, are considered a substitute for the former. The sacrifice of animals in temples, or the saying of masses, the erection of chapels or crosses by the roadside, are soon regarded as the most meritorious works; so that even a great crime may be expiated by them, as also by penance, subjection to priestly authority, confessions, pilgrimages, donations to the temple and its priests, the building of monasteries and the like; until finally the clergy appear almost only as mediators in the corruption of the gods. And if things do not go so far as that, where is the religion whose confessors do not consider prayers, songs of praise, and various kinds of devotional exercise, at any rate, a partial substitute for moral conduct? Look at England, for instance, where the audacious priestcraft has mendaciously identified the Christian Sunday with the Jewish Sabbath, in spite of the fact that it was ordained by

Constantine the Great in opposition to the Jewish Sabbath, and even took its name, so that Jehovah's ordinances for the Sabbath— *i.e.*, the day on which the Almighty rested, tired after His six days' work, making it therefore *essentially the last day* of the week—might be conferred on the Christian Sunday, the *dies solis*, the first day of the week which the sun opens in glory, the day of devotion and joy. The result of this fraud is that in England "Sabbath breaking," or the "desecration of the Sabbath," that is, the slightest occupation, whether it be of a useful or pleasurable nature, and any kind of game, music, knitting, or worldly book, are on Sundays regarded as great sins. Must not the ordinary man believe that if, as his spiritual guides impress upon him, he never fails in a "strict observance of the holy Sabbath and a regular attendance on Divine Service,"—in other words, if he invariably whiles away his time on a Sunday, and never fails to sit two hours in church to listen to the same Litany for the thousandth time, and to babble it with the rest *a tempo*, he may reckon on indulgence in here and there little sins which he at times allows himself? Those devils in human form, the slave-owners and slave-traders in the Free States of North America (they should be called the Slave States), are, in general, orthodox, pious Anglicans, who look upon it as a great sin to work on Sundays; and confident in this, and their regular attendance at church, they expect to gain eternal happiness. The demoralising influence of religion is less problematical than its moral influence. On the other hand, how great and how certain that moral influence must be to make amends for the horrors and misery which religions, especially the Christian and Mohammedan religions, have occasioned and spread over the earth! Think of the fanaticism, of the endless persecutions, the religious wars, that sanguinary frenzy of which the ancients had no idea; then, think of the Crusades, a massacre lasting two hundred years, and perfectly unwarrantable, with its war-cry, *It is God's will*, so that it might get into its possession the grave of one who had preached love and endurance; think of the cruel expulsion and extermination of the Moors and Jews from Spain; think of the massacres, of the inquisitions and other heretical tribunals, the bloody and terrible conquests of the Mohammedans in three different parts of the world, and the conquest of the Christians in America, whose inhabitants were for the most part, and in Cuba entirely, exterminated; according to Las Casas, within forty years twelve million persons were murdered—of course, all *in majorem Dei gloriam,* and for the spreading of the Gospel, and because, moreover, what was not Christian was not looked upon as human. It is true I have already touched upon these matters; but when in our day "the Latest News from the Kingdom of God" is printed, we shall not be tired of bringing older news to mind. And in particular, let us not forget India, that sacred soil, that cradle of the human race, at any rate of the race to which we belong, where first Mohammedans, and later Christians, were most cruelly infuriated against the followers of the original belief of mankind; and the eternally lamentable, wanton, and cruel destruction and disfigurement of the most ancient temples and images, still show traces of the monotheistic rage of the Mohammedans, as it was carried on from Marmud the Ghaznevid of accursed memory, down to Aureng Zeb, the fratricide, whom later the Portuguese Christians faithfully tried to imitate by destroying the temples and the *auto da fe* of the inquisition at Goa. Let us also not forget the chosen people of God, who, after they had, by Jehovah's express and special command, stolen from their old and faithful friends in Egypt the gold and silver vessels which had been lent to them, made a murderous and predatory excursion into the Promised Land, with Moses at their head, in order to tear it from the rightful owners, also at Jehovah's express and repeated commands, knowing no compassion, and relentlessly murdering and exterminating all the inhabitants, even the women and children (Joshua x., xi.); just

145

because they were not circumcised and did not know Jehovah, which was sufficient reason to justify every act of cruelty against them. For the same reason, in former times the infamous roguery of the patriarch Jacob and his chosen people against Hamor, King of Shalem, and his people is recounted to us with glory, precisely because the people were unbelievers. Truly, it is the worst side of religions that the believers of one religion consider themselves allowed everything against the sins of every other, and consequently treat them with the utmost viciousness and cruelty; the Mohammedans against the Christians and Hindoos; the Christians against the Hindoos, Mohammedans, Americans, Negroes, Jews, heretics, and the like. Perhaps I go too far when I say *all* religions; for in compliance with truth, I must add that the fanatical horrors, arising from religion, are only perpetrated by the followers of the monotheistic religions, that is, of Judaism and its two branches, Christianity and Islamism. The same is not reported of the Hindoos and Buddhists, although we know, for instance, that Buddhism was driven out about the fifth century of our era by the Brahmans from its original home in the southernmost part of the Indian peninsula, and afterwards spread over the whole of Asia; yet we have, so far as I know, no definite information of any deeds of violence, of wars and cruelties by which this was brought about. This may, most certainly, be ascribed to the obscurity in which the history of those countries is veiled; but the extremely mild character of their religion, which continually impresses upon us to be forbearing towards *every living thing*, as well as the circumstance that Brahmanism properly admits no proselytes by reason of its caste system, leads us to hope that its followers may consider themselves exempt from shedding blood to any great extent, and from cruelty in any form. Spence Hardy, in his excellent book on *Eastern Monachism*, p. 412, extols the extraordinary tolerance of the Buddhists, and adds his assurance that the annals of Buddhism furnish fewer examples of religious persecution than those of any other religion. As a matter of fact, intolerance is only essential to monotheism: an only god is by his nature a jealous god, who cannot permit any other god to exist. On the other hand, polytheistic gods are by their nature tolerant: they live and let live; they willingly tolerate their colleagues as being gods of the same religion, and this tolerance is afterwards extended to alien gods, who are, accordingly, hospitably received, and later on sometimes attain even the same rights and privileges; as in the case of the Romans, who willingly accepted and venerated Phrygian, Egyptian, and other foreign gods. Hence it is the monotheistic religions alone that furnish us with religious wars, persecutions, and heretical tribunals, and also with the breaking of images, the destruction of idols of the gods; the overthrowing of Indian temples and Egyptian colossi, which had looked on the sun three thousand years; and all this because a jealous God had said: *"Thou shalt make no graven image,"* etc. To return to the principal part of the matter: you are certainly right in advocating the strong metaphysical needs of mankind; but religions appear to me to be not so much a satisfaction as an abuse of those needs. At any rate we have seen that, in view of the progress of morality, its advantages are for the most part problematical, while its disadvantages, and especially the enormities which have appeared in its train, are obvious. Of course the matter becomes quite different if we consider the utility of religion as a mainstay of thrones; for in so far as these are bestowed "by the grace of God," altar and throne are closely related. Accordingly, every wise prince who loves his throne and his family will walk before his people as a type of true religion; just as even Machiavelli, in the eighteenth chapter of his book, urgently recommended religion to princes. Moreover, it may be added that revealed religions are related to philosophy, exactly as the sovereigns by the grace of

God are to the sovereignty of the people; and hence the two former terms of the parallel are in natural alliance.

Demop. Oh, don't adopt that tone! But consider that in doing so you are blowing the trumpet of ochlocracy and anarchy, the arch-enemy of all legislative order, all civilisation, and all humanity.

Phil. You are right. It was only a sophism, or what the fencing-master calls a feint. I withdraw it therefore. But see how disputing can make even honest men unjust and malicious. So let us cease.

Demop. It is true I regret, after all the trouble I have taken, that I have not altered your opinion in regard to religion; on the other hand, I can assure you that everything you have brought forward has not shaken my conviction of its high value and necessity.

Phil. I believe you; for as it is put in Hudibras:

> "He that complies against his will
> Is of his own opinion still."

I find consolation, however, in the fact that in controversies and in taking mineral waters, it is the after-effects that are the true ones.

Demop. I hope the after-effect may prove to be beneficial in your case.

Phil. That might be so if I could only digest a Spanish proverb.

Demop. And that is?

Phil. Detras de la cruz esta el Diablo.

Demop. Which means?

Phil Wait—"Behind the cross stands the devil."

Demop. Come, don't let us separate from each other with sarcasms, but rather let us allow that religion, like Janus, or, better still, like the Brahman god of death, Yama, has two faces, and like him, one very friendly and one very sullen. Each of us, however, has only fixed his eyes on one.

Phil. You are right, old fellow.

FOOTNOTES:

13   *De Anim. Mundi,* p. 104, d. Steph.

PSYCHOLOGICAL OBSERVATIONS.

Every animal, and especially man, requires, in order to exist and get on in the world, a certain fitness and proportion between his will and his intellect. The more exact and true this fitness and proportion are by nature, the easier, safer, and pleasanter it will be for him to get through the world. At the same time, a mere approximation to this exact point will protect him from destruction. There is, in consequence, a certain scope within the limits of exactness and fitness of this so-called proportion. The normal proportion is as follows. As the object of the intellect is to be the light and guide of the will on its path, the more violent, impetuous, and passionate the inner force of the will, the more perfect and clear must be the intellect which belongs to it; so that the ardent efforts of the will, the glow of passion, the vehemence of affection, may not lead a man astray or drive him to do things that he has not given his consideration or are wrong or will ruin him; which will infallibly be the case when a very strong will is combined with a very weak intellect. On the other hand, a phlegmatic character, that is to say, a weak and feeble will, can agree and get on with little intellect; a moderate will only requires a moderate intellect. In general, any disproportion between the will and intellect—that is to say, any deviation from the normal proportion referred to—tends to make a man unhappy; and the same thing happens when the disproportion is reversed. The development of the intellect to an abnormal degree of strength and superiority, thereby making it out of all proportion to the will, a condition which constitutes the essence of true genius, is not only superfluous but actually an impediment to the needs and purposes of life. This means that, in youth, excessive energy in grasping the objective world, accompanied by a lively imagination and little experience, makes the mind susceptible to exaggerated ideas and a prey even to chimeras; and this results in an eccentric and even fantastic character. And when, later, this condition of mind no longer exists and succumbs to the teaching of experience, the genius will never feel so much at home or take up his position in the everyday world or in civic life, and move with the ease of a man of normal intellect; indeed, he is often more apt to make curious mistakes. For the ordinary mind is so perfectly at home in the narrow circle of its own ideas and way of grasping things that no one can control it in that circle; its capacities always remain true to their original purpose, namely, to look after the service of the will; therefore it applies itself unceasingly to this end without ever going beyond it. While the genius, as I have stated, is at bottom a *monstrum per excessum*; just as conversely the passionate, violent, and unintelligent man, the brainless savage, is a *monstrum per dejectum*.

* * * * *

The *will* to *live*, which forms the innermost kernel of every living being, is most distinctly apparent in the highest, that is to say in the cleverest, order of animals, and therefore in them we may see and consider the nature of the will most clearly. For *below* this order of animals the will is not so prominent, and has a less degree of objectivation; but *above* the higher order of animals, I mean in men, we get reason, and with reason reflection, and with this the faculty for dissimulation, which immediately throws a veil over the actions of the will. But in outbursts of affection and passion the will exhibits itself unveiled. This is precisely why passion, when it speaks, always carries conviction, whatever the passion may be; and rightly so. For the same reason, the passions are the principal theme of poets and the stalking-horse of actors. And it is because the will is most striking in the lower class of animals that we may account for our delight in dogs, apes, cats, etc.; it is the absolute *naivete* of all their expressions which charms us so much.

What a peculiar pleasure it affords us to see any free animal looking after its own welfare unhindered, finding its food, or taking care of its young, or associating with others of its kind, and so on! This is exactly what ought to be and can be. Be it only a bird, I can look at it for some time with a feeling of pleasure; nay, a water-rat or a frog, and with still greater pleasure a hedgehog, a weazel, a roe, or a deer. The contemplation of animals delights us so much, principally because we see in them our own existence very *much simplified*.

There is only one mendacious creature in the world—man. Every other is true and genuine, for it shows itself as it is, and expresses itself just as it feels. An emblematical or allegorical expression of this fundamental difference is to be found in the fact that all animals go about in their natural state; this largely accounts for the happy impression they make on us when we look at them; and as far as I myself am concerned, my heart always goes out to them, particularly if they are free animals. Man, on the other hand, by his silly dress becomes a monster; his very appearance is objectionable, enhanced by the unnatural paleness of his complexion,—the nauseating effect of his eating meat, of his drinking alcohol, his smoking, dissoluteness, and ailments. He stands out as a blot on Nature. And it was because the Greeks were conscious of this that they restricted themselves as far as possible in the matter of dress.

* * * * *

Much that is attributed to *force of habit* ought rather to be put down to the constancy and immutability of original, innate character, whereby we always do the *same* thing under the same circumstances; which happens the first as for the hundredth time in consequence of the same necessity. While *force of habit*, in reality, is solely due to *indolence* seeking to save the intellect and will the work, difficulty, and danger of making a fresh choice; so that we are made to do to-day what we did yesterday and have done a hundred times before, and of which we know that it will gain its end.

But the truth of the matter lies deeper; for it can be explained more clearly than appears at first sight. The *power of inertia* applied to bodies which may be moved by mechanical means only, becomes *force of habit* when applied to bodies which are moved by motives. The actions which we do out of sheer force of habit occur, as a matter of fact, without any individual separate motive exercised for the particular case; hence we do not really think of them. It was only when each action at first took place that it had a motive; after that it became a habit; the secondary after-effect of this motive is the present habit, which is sufficient to carry on the action; just as a body, set in motion by a push, does not need another push in order to enable it to continue its motion; it will continue in motion for ever if it is not obstructed in any way. The same thing applies to animals; training is a habit which is forced upon them. The horse draws a cart along contentedly without being urged to do so; this motion is still the effect of those lashes with the whip which incited him at first, but which by the law of inertia have become perpetuated as habit. There is really something more in all this than a mere parable; it is the identity of the thing in question, that is to say of the will, at very different degrees of its objectivation, by which the same law of motion takes such different forms.

* * * * *

Viva muchos anos! is the ordinary greeting in Spain, and it is usual throughout the whole world to wish people a long life. It is not a knowledge of what life is that explains the origin of such a wish, but rather knowledge of what man is in his real nature: namely, *the will to live*.

The wish which every one has, that he may be *remembered* after his death, and which those people with aspirations have for *posthumous* fame, seems to me to arise from this tenacity to life. When they see themselves cut off from every possibility of real existence they struggle after a life which is still within their reach, even if it is only an ideal—that is to say, an unreal one.

* * * * *

We wish, more or less, to get to the end of everything we are interested in or occupied with; we are impatient to get to the end of it, and glad when it is finished. It is only the general end, the end of all ends, that we wish, as a rule, as far off as possible.

* * * * *

Every separation gives a foretaste of death, and every meeting a foretaste of the resurrection. This explains why even people who were indifferent to each other, rejoice so much when they meet again after the lapse of twenty or thirty years.

* * * * *

The deep sorrow we feel on the death of a friend springs from the feeling that in every individual there is a something which we cannot define, which is his alone and therefore *irreparable. Omne individuum ineffabile.* The same applies to individual animals. A man who has by accident fatally wounded a favourite animal feels the most acute sorrow, and the animal's dying look causes him infinite pain.

* * * * *

It is possible for us to grieve over the death of our enemies and adversaries, even after the lapse of a long time, almost as much as over the death of our friends—that is to say, if we miss them as witnesses of our brilliant success.

* * * * *

That the sudden announcement of some good fortune may easily have a fatal effect on us is due to the fact that our happiness and unhappiness depend upon the relation of our demands to what we get; accordingly, the good things we possess, or are quite sure of possessing, are not felt to be such, because the nature of all enjoyment is really only *negative*, and has only the effect of annulling pain; whilst, on the other hand, the nature of pain or evil is really positive and felt immediately. With the possession, or the certain prospect of it, our demands instantly rise and increase our desire for further possession and greater prospects. But if the mind is depressed by continual misfortune, and the claims reduced to a *minimum*, good fortune that comes suddenly finds no capacity for its acceptance. Neutralised by no previous claims, it now has apparently a positive effect, and accordingly its whole power is exercised; hence it may disorganise the mind—that is to say, be fatal to it. This is why, as is well known, one is so careful to get a man first to hope for happiness before announcing it, then to suggest the prospect of it, then little by little make it known, until gradually all is known to him; every portion of the revelation loses the strength of its effect because it is anticipated by a demand, and room is still left for more. In virtue of all this, it might be said that our stomach for good fortune is bottomless, but the entrance to it is narrow. What has been said does not apply to sudden misfortunes in the same way. Since hope always resists them, they are for this reason rarely fatal. That fear does not perform an analogous office in cases of good fortune is due to the fact that we are instinctively more inclined to hope than to fear; just as our eyes turn of themselves to light in preference to darkness.

* * * * *

Hope is to confuse the desire that something should occur with the probability that it will. Perhaps no man is free from this folly of the heart, which deranges the intellect's correct estimation of probability to such a degree as to make him think the event quite possible, even if the chances are only a thousand to one. And still, an unexpected misfortune is like a speedy death-stroke; while a hope that is always frustrated, and yet springs into life again, is like death by slow torture.

He who has given up hope has also given up fear; this is the meaning of the expression *desperate*. It is natural for a man to have faith in what he wishes, and to have faith in it because he wishes it. If this peculiarity of his nature, which is both beneficial and comforting, is eradicated by repeated hard blows of fate, and he is brought to a converse condition, when he believes that something must happen because he does not wish it, and what he wishes can never happen just because he wishes it; this is, in reality, the state which has been called *desperation*.

*　*　*　*　*

That we are so often mistaken in others is not always precisely due to our faulty judgment, but springs, as a rule as Bacon says, from *intellectus luminis sicci non est, sec recipit infusionem a voluntate et affectibus*: for without knowing it, we are influenced for or against them by trifles from the very beginning. It also often lies in the fact that we do not adhere to the qualities which we really discover in them, but conclude from these that there are others which we consider inseparable from, or at any rate incompatible with, them. For instance, when we discern generosity, we conclude there is honesty; from lying we conclude there is deception; from deception, stealing, and so on; and this opens the door to many errors, partly because of the peculiarity of human nature, and partly because of the one-sidedness of our point of view. It is true that character is always consistent and connected; but the roots of all its qualities lies too deep to enable one to decide from special data in a given case which qualities can, and which cannot exist together.

*　*　*　*　*

The use of the word *person* in every European language to signify a human individual is unintentionally appropriate; *persona* really means a player's mask, and it is quite certain that no one shows himself as he is, but that each wears a mask and plays a *role*. In general, the whole of social life is a continual comedy, which the worthy find insipid, whilst the stupid delight in it greatly.

* * * * *

It often happens that we blurt out things that may in some kind of way be harmful to us, but we are silent about things that may make us look ridiculous; because in this case effect follows very quickly on cause.

* * * * *

The ordinary man who has suffered injustice burns with a desire for revenge; and it has often been said that revenge is sweet. This is confirmed by the many sacrifices made merely for the sake of enjoying revenge, without any intention of making good the injury that one has suffered. The centaur Nessus utilised his last moments in devising an extremely clever revenge, and the fact that it was certain to be effective sweetened an otherwise bitter death. The same idea, presented in a more modern and plausible way, occurs in Bertolotti's novel, *Le due Sorelle* which has been translated into three languages. Walter Scott expresses mankind's proneness to revenge in words as powerful as they are true: "Vengeance is the sweetest morsel to the mouth that ever was cooked in hell!" I shall now attempt a psychological explanation of revenge. All the suffering that nature, chance, or fate have assigned to us does not, *ceteris paribus*, pain us so much as suffering which is brought upon us by the arbitrary will of another. This is due to the fact that we regard nature and fate as the original rulers of the world; we look upon what befalls us, through them, as something that might have befallen every one else. Therefore in a case of suffering which arises from this source,

157

we bemoan the fate of mankind in general more than we do our own. On the other hand, suffering inflicted on us through the arbitrary will of another is a peculiarly bitter addition to the pain or injury caused, as it involves the consciousness of another's superiority, whether it be in strength or cunning, as opposed to our own weakness. If compensation is possible, it wipes out the injury; but that bitter addition, "I must submit to that from you," which often hurts more than the injury itself, is only to be neutralised by vengeance. For by injuring the man who has injured us, whether it be by force or cunning, we show our superiority, and thereby annul the proof of his. This gives that satisfaction to the mind for which it has been thirsting. Accordingly, where there is much pride or vanity there will be a great desire for revenge. But as the fulfilment of every wish proves to be more or less a delusion, so is also the wish for revenge. The expected enjoyment is mostly embittered by pity; nay, gratified revenge will often lacerate the heart and torment the mind, for the motive which prompts the feeling of it is no longer active, and what is left is the testimony of our wickedness.

* * * * *

The pain of an ungratified desire is small compared with that of repentance; for the former has to face the immeasurable, open future; the latter the past, which is closed irrevocably.

* * * * *

Money is human happiness *in abstracto*; so that a man who is no longer capable of enjoying it *in concrete* gives up his whole heart to it.

* * * * *

Moroseness and melancholy are very opposite in nature; and melancholy is more nearly related to happiness than to moroseness.

158

Melancholy attracts; moroseness repels. Hypochondria not only makes us unreasonably cross and angry over things concerning the present; not only fills us with groundless fears of imaginative mishaps for the future; but also causes us to unjustly reproach ourselves concerning our actions in the past.

Hypochondria causes a man to be always searching for and racking his brain about things that either irritate or torment him. The cause of it is an internal morbid depression, combined often with an inward restlessness which is temperamental; when both are developed to their utmost, suicide is the result.

* * * * *

What makes a man hard-hearted is this, that each man has, or fancies he has, sufficient in his own troubles to bear. This is why people placed in happier circumstances than they have been used to are sympathetic and charitable. But people who have always been placed in happy circumstances are often the reverse; they have become so estranged to suffering that they have no longer any sympathy with it; and hence it happens that the poor sometimes show themselves more benevolent than the rich.

On the other hand, what makes a man so very *curious*, as may be seen in the way he will spy into other people's affairs, is boredom, a condition which is diametrically opposed to suffering;—though envy also often helps in creating curiosity.

* * * * *

At times, it seems as though we wish for something, and at the same time do not wish for it, so that we are at once both pleased and troubled about it. For instance, if we have to undergo some decisive test in some affair or other, in which to come off victorious is of great importance to us; we both wish that the time to be tested were here, and yet dread the idea of its coming. If it happens that the time, for once in a way, is

159

postponed, we are both pleased and sorry, for although the postponement was unexpected, it, however, gives us momentary relief. We have the same kind of feeling when we expect an important letter containing some decision of moment, and it fails to come.

In cases like these we are really controlled by two different motives; the stronger but more remote being the desire to stand the test, and to have the decision given in our favour; the weaker, which is closer at hand, the desire to be left in peace and undisturbed for the present, and consequently in further enjoyment of the advantage that hoping on in uncertainty has over what might possibly be an unhappy issue. Consequently, in this case the same happens to our moral vision as to our physical, when a smaller object near at hand conceals from view a bigger object some distance away.

* * * * *

The course and affairs of our individual life, in view of their true meaning and connection, are like a piece of crude work in mosaic. So long as one stands close in front of it, one cannot correctly see the objects presented, or perceive their importance and beauty; it is only by standing some distance away that both come into view. And in the same way one often understands the true connection of important events in one's own life, not while they are happening, or even immediately after they have happened, but only a long time afterwards.

Is this so, because we require the magnifying power of imagination, or because a general view can only be got by looking from a distance? or because one's emotions would otherwise carry one away? or because it is only the school of experience that ripens our judgment? Perhaps all these combined. But it is certain that it is only after many years that we see the actions of others, and sometimes even our own, in their true light. And as it is in one's own life, so it is in history.

* * * * *

Why is it, in spite of all the mirrors in existence, no man really knows what he looks like, and, therefore, cannot picture in his mind his own person as he pictures that of an acquaintance? This is a difficulty which is thwarted at the very outset by *gnothi sauton*—*know thyself*.

This is undoubtedly partly due to the fact that a man can only see himself in the glass by looking straight towards it and remaining quite still; whereby the play of the eye, which is so important, and the real characteristic of the face is, to a great extent, lost. But co-operating with this physical impossibility, there appears to be an ethical impossibility analogous to it. A man cannot regard the reflection of his own face in the glass as if it were the face of *some one else*—which is the condition of his seeing himself *objectively*. This objective view rests with a profound feeling on the egoist's part, as a moral being, that what he is looking at is *not himself*, which is requisite for his perceiving all his defects as they really are from a purely objective point of view; and not until, then can he see his face reflected as it really and truly is. Instead of that, when a man sees his own person in the glass the egoistic side of him always whispers, *It is not somebody else, but I myself*, which has the effect of a *noli me tangere*, and prevents his taking a purely objective view. Without the leaven of a grain of malice, it does not seem possible to look at oneself objectively.

* * * * *

No one knows what capacities he possesses for suffering and doing until an opportunity occurs to bring them into play; any more than he imagines when looking into a perfectly smooth pond with a mirror-like surface, that it can tumble and toss and rush from rock to rock, or leap as high into the air as a fountain;—any more than in ice-cold water he suspects latent warmth.

* * * * *

161

That line of Ovid's,

"Pronaque cum spectent animalia cetera terram,"

is only applicable in its true physical sense to animals; but in a figurative and spiritual sense, unfortunately, to the great majority of men too. Their thoughts and aspirations are entirely devoted to physical enjoyment and physical welfare, or to various personal interests which receive their importance from their relation to the former; but they have no interests beyond these. This is not only shown in their way of living and speaking, but also in their look, the expression of their physiognomy, their gait and gesticulations; everything about them proclaims *in terram prona!* Consequently it is not to them, but only to those nobler and more highly endowed natures, those men who really think and observe things round them, and are the exceptions in the human race, that the following lines are applicable:

"Os homini sublime dedit coelumque tueri
Jussitt et erectos ad sidera tollere vultus."

* * * * *

Why is *"common"* an expression of contempt? And why are *"uncommon," "extraordinary," "distinguished,"* expressions of approbation? Why is everything that is common contemptible?

Common, in its original sense, means that which is peculiar and common to the whole species, that is to say that which is innate in the species. Accordingly, a man who has no more qualities than those of the human species in general is a *"common man" "Ordinary man"* is a much milder expression, and is used more in reference to what is intellectual, while *common* is used more in a moral sense.

162

What value can a being have that is nothing more than like millions of its kind? Millions? Nay, an infinitude, an endless number of beings, which Nature in *secula seculorum* unceasingly sends bubbling forth from her inexhaustible source; as generous with them as the smith with the dross that flies round his anvil.

So it is evidently only right that a being which has no other qualities than those of the species, should make no claim to any other existence than that confined to and conditioned by the species.

I have already several times explained[14] that whilst animals have only the generic character, it falls to man's share alone to have an individual character. Nevertheless, in most men there is in reality very little individual character; and they may be almost all classified. *Ce sont des especes.* Their desires and thoughts, like their faces, are those of the whole species—at any rate, those of the class of men to which they belong, and they are therefore of a trivial, common nature, and exist in thousands. Moreover, as a rule one can tell pretty exactly beforehand what they will say and do. They have no individual stamp: they are like manufactured goods. If, then, their nature is absorbed in that of the species, must not their existence be too? The curse of vulgarity reduces man to the level of animals, for his nature and existence are merged in that of the species only. It is taken for granted that anything that is high, great, or noble by its very nature stands isolated in a world where no better expression can be found to signify what is base and paltry than the term which I have mentioned as being generally used—namely, *common.*

* * * * *

According as our intellectual energy is strained or relaxed will life appear to us either so short, petty, and fleeting, that nothing can happen of sufficient importance to affect our feelings; nothing is of any importance to us—be it pleasure, riches, or even fame, and however much we may have failed, we cannot have lost much; or *vice versa,* life will appear so long, so important, so all in all, so grave, and so difficult that we throw ourselves into it with our whole soul, so that we may get a share of its possessions, make ourselves sure of its prizes, and carry out our plans. The latter is the immanent view of life; it is what Gracian means by his expression, *tomar muy de veras el vivir* (life is to be taken seriously); while for the former, the transcendental view, Ovid's *non est tanti* is a good expression; Plato's a still better, [Greek: oute ti ton anthropinon axion hesti, megalaes spoudaes] *(nihil, in rebus humanis, magno studio dignum est)*.

The former state of mind is the result of the intellect having gained ascendency over consciousness, where, freed from the mere service of the will, it grasps the phenomena of life objectively, and so cannot fail to see clearly the emptiness and futility of it. On the other hand, it is the *will* that rules in the other condition of mind, and it is only there to lighten the way to the object of its desires. A man is great or small according to the predominance of one or the other of these views of life.

* * * * *

It is quite certain that many a man owes his life's happiness solely to the circumstance that he possesses a pleasant smile, and so wins the hearts of others. However, these hearts would do better to take care to remember what Hamlet put down in his tablets—*that one may smile, and smile, and be a villain.*

* * * * *

People of great and brilliant capacities think little of admitting or exposing their faults and weaknesses. They regard them as something for which they have paid, and even are of the opinion that these weaknesses, instead of being a disgrace to them, do them honour. This is especially the case when they are errors that are inseparable from their brilliant capacities—*conditiones sine quibus non*, or, as George Sand expressed it, *chacun a les defauts de ses vertus.*

On the contrary, there are people of good character and irreproachable minds, who, rather than admit their few little weaknesses, carefully conceal them, and are very sensitive if any reference is made to them; and this just because their whole merit consists in the absence of errors and defects; and hence when these errors come to light they are immediately held in less esteem.

* * * * *

Modesty, in people of moderate ability, is merely honesty, but in people of great talent it is hypocrisy. Hence it is just as becoming in the latter to openly admit the regard they have for themselves, and not to conceal the fact that they are conscious of possessing exceptional capabilities, as it is in the former to be modest. Valerius Maximus gives some very good examples of this in his chapter *de fiducia sui*.

* * * * *

Man even surpasses all the lower order of animals in his capacity for being trained. Mohammedans are trained to pray five times a day with their faces turned towards Mecca; and they do it regularly. Christians are trained to make the sign of the Cross on certain occasions, and to bow, and so forth; so that religion on the whole is a real masterpiece of training—that is to say, it trains people what they are to think; and the training, as is well known, cannot begin too early. There is no absurdity, however palpable it may be, which may not be fixed in the minds of all men, if it is inculcated before they are six years old by continual and earnest repetition. For it is the same with men as with animals, to train them with perfect success one must begin when they are very young.

Noblemen are trained to regard nothing more sacred than their word of honour, to believe earnestly, rigidly, and firmly in the inane code of knight-errantry, and if necessary to seal their belief by death, and to look upon a king as a being of a higher order. Politeness and compliments, and particularly our courteous attitude towards ladies, are the result of training; and so is our esteem for birth, position, and title. And so is our displeasure at certain expressions directed against us, our displeasure being proportionate to the expression used. The Englishman has been trained to consider his being called no gentleman a crime worthy of death—a liar, a still greater crime; and so, the Frenchman, if he is called a coward; a German, if he is called a stupid. Many people are trained to be honest in some particular direction, whilst in everything else they exhibit very little honesty; so that many a man will not steal money, but he will steal everything that will afford him enjoyment in an indirect way. Many a shopkeeper will deceive without scruple, but he will on no condition whatever steal.

* * * * *

The doctor sees mankind in all its weakness; the lawyer in all its wickedness; the theologian in all its stupidity.

* * * * *

Opinion obeys the same law as the swing of the pendulum: if it goes beyond the centre of gravity on one side, it must go as far beyond on the other. It is only after a time that it finds the true point of rest and remains stationary.

* * * * *

Distance in space decreases the size of things, for it contracts them and so makes their defects and deficiencies disappear. This is why everything looks so much finer in a contracting mirror or in a *camera obscura* than it is in reality; and the past is affected in the same way in the course of time. The scenes and events that happened long ago, as well as the persons who took part in them, become a delight to the memory, which ignores everything that is immaterial and disagreeable. The present possesses no such advantage; it always seems to be defective. And in space, small objects near at hand appear to be big, and if they are very near, they cover the whole of our field of vision; but as soon as we stand some little distance away they become minute and finally invisible. And so it is with time: the little affairs and misfortunes of everyday life excite in us emotion, anxiety, vexation, passion, for so long as they are quite near us, they appear big, important, and considerable; but as soon as the inexhaustible stream of time has carried them into the distance they become unimportant; they are not worth remembering and are soon forgotten, because their importance merely consisted in being near.

* * * * *

It is only now and then that a man learns something; but he forgets the whole day long.

Our memory is like a sieve, that with time and use holds less and less; in so far, namely, as the older we get, the quicker anything we have entrusted to our memory slips through it, while anything that was fixed firmly in it, when we were young, remains. This is why an old man's recollections are the clearer the further they go back, and the less clear the nearer they approach the present; so that his memory, like his eyes, becomes long-sighted ([Greek: presbus]).

That sometimes, and apparently without any reason, long-forgotten scenes suddenly come into the memory, is, in many cases, due to the recurrence of a scarcely perceptible odour, of which we were conscious when those scenes actually took place; for it is well known that odours more easily than anything else awaken memories, and that, in general, something of an extremely trifling nature is all that is necessary to call up a *nexus idearum*.

And by the way, I may say that the sense of sight has to do with the understanding,[15] the sense of hearing with reason,[16] and the sense of smell with memory, as we see in the present case. Touch and taste are something real, and dependent on contact; they have no ideal side.

* * * * *

Memory has also this peculiarity attached to it, that a slight state of intoxication very often enhances the remembrance of past times and scenes, whereby all the circumstances connected with them are recalled more distinctly than they could be in a state of sobriety; on the other hand, the recollection of what one said or did while in a state of intoxication is less clear than usual, nay, one does not recollect at all if one has been very drunk. Therefore, intoxication enhances one's recollection of the past, while, on the other hand, one remembers little of the present, while in that state.

* * * * *

That arithmetic is the basest of all mental activities is proved by the fact that it is the only one that can be accomplished by means of a machine. Take, for instance, the reckoning machines that are so commonly used in England at the present time, and solely for the sake of convenience. But all *analysis finitorum et infinitorum* is fundamentally based on calculation. Therefore we may gauge the "profound sense of the mathematician," of whom Lichtenberg has made fun, in that he says: "These so-called professors of mathematics have taken advantage of the ingenuousness of other people, have attained the credit of possessing profound sense, which strongly resembles the theologians' profound sense of their own holiness."

* * * * *

As a rule, people of very great capacities will get on better with a man of extremely limited intelligence than with a man of ordinary intelligence;

and it is for the same reason that the despot and the plebeians, the grandparents and the grandchildren, are natural allies.

* * * * *

I am not surprised that people are bored when they are alone; they cannot laugh when they are alone, for such a thing seems foolish to them. Is laughter, then, to be regarded as merely a signal for others, a mere sign, like a word? It is a want of imagination and dulness of mind generally ([Greek: anaisthaesia kai bradytaes psychaes]), as Theophrastus puts it, that prevents people from laughing when they are alone. The lower animals neither laugh when they are alone nor in company.

Nyson, the misanthropist, was surprised as he was laughing to himself by one of these people, who asked him why he laughed when he was alone. "That is just why I was laughing," was the answer.

* * * * *

People who do not go to the theatre are like those who make their toilet without a looking-glass;—but it is still worse to come to a decision without seeking the advice of a friend. For a man may have the most correct and excellent judgment in everything else but in his own affairs; because here the will at once deranges the intellect. Therefore a man should seek counsel. A doctor can cure every one but himself; this is why he calls in a colleague when he is ill.

* * * * *

The natural gesticulation of everyday life, such as accompanies any kind of lively conversation, is a language of its own, and, moreover, is much more universal than the language of words; so far as it is independent of words, and the same in all nations; although each nation makes use of gesticulation in proportion to its vivacity, and in individual nations, the

Italian, for instance, it is supplemented by some few gesticulations which are merely conventional, and have therefore only local value.

Its universal use is analogous to logic and grammar, since it expresses the form and not the matter of conversation. However, it is to be distinguished from them since it has not only an intellectual relation but also a moral—that is, it defines the movements of the will. And so it accompanies conversation, just as a correctly progressive bass accompanies a melody, and serves in the same way to enhance the effect. The most interesting fact about gesticulation is that as soon as conversation assumes the same *form* there is a repetition of the same gesture. This is the case, however varied the *matter*, that is to say, the subject-matter, may be. So that I am able to understand quite well the general nature of a conversation—in other words, the mere form and type of it, while looking out of a window—without hearing a word spoken. It is unmistakably evident that the speaker is arguing, advancing his reasons, then modifying them, then urging them, and drawing his conclusion in triumph; or it may be he is relating some wrong that he has suffered, plainly depicting in strong and condemnatory language the stupidity and stubbornness of his opponents; or he is speaking of the splendid plan he has thought out and put in execution, explaining how it became a success, or perhaps failed because fate was unfavourable; or perhaps he is confessing that he was powerless to act in the matter in question; or recounting that he noticed and saw through, in good time, the evil schemes that had been organised against him, and by asserting his rights or using force frustrated them and punished their author; and a hundred other things of a similar kind. But what gesticulation alone really conveys to me is the essential matter—be it of a moral or intellectual nature— of the whole conversation *in abstracto*. That is to say the quintessence, the true substance of the conversation, remains identical whatever has brought about the conversation, and consequently whatever the subject-matter of it may be.

The most interesting and amusing part of the matter, as has been said, is the complete identity of the gestures for denoting the same kind of circumstances, even if they are used by most diverse people; just as the words of a language are alike for every one and liable to such modifications as are brought about by a slight difference in accent or education. And yet these standing forms of gesticulation which are universally observed are certainly the outcome of no convention; they are natural and original, a true language of nature, which may have been strengthened by imitation and custom. It is incumbent on an actor, as is well known, and on a public speaker, to a less extent, to make a careful study of gesture—a study which must principally consist in the observation and imitation of others, for the matter cannot very well be based on abstract rules; with the exception of some quite general leading principles—as, for instance, that the gesture must not follow the word, but rather immediately precede it, in order to announce it and thereby rouse attention.

The English have a peculiar contempt for gesticulation, and regard it as something undignified and common; this seems to me to be only one of those silly prejudices of English fastidiousness. For it is a language which nature has given to every one and which every one understands; therefore to abolish and forbid it for no other reason than to gratify that so much extolled, gentlemanly feeling, is a very dubious thing to do.

* * * * *

The state of human happiness, for the most part, is like certain groups of trees, which seen from a distance look wonderfully fine; but if we go up to them and among them, their beauty disappears; we do not know wherein it lay, for it is only trees that surround us. And so it happens that we often envy the position of others.

FOOTNOTES:

14 *Grundpr. der Ethik*, p. 48; *Welt als Wille und Vorstellung*, vol. i. p. 338.
15 *Vierfache Wurzel*, Sec. 21.
16 *Pererga*, vol. ii. Sec. 311.

METAPHYSICS OF LOVE.

We are accustomed to see poets principally occupied with describing the love of the sexes. This, as a rule, is the leading idea of every dramatic work, be it tragic or comic, romantic or classic, Indian or European. It in no less degree constitutes the greater part of both lyric and epic poetry, especially if in these we include the host of romances which have been produced every year for centuries in every civilised country in Europe as regularly as the fruits of the earth. All these works are nothing more than many-sided, short, or long descriptions of the passion in question. Moreover, the most successful delineations of love, such, for example, as *Romeo and Juliet*, *La Nouvelle Heloise*, and *Werther*, have attained immortal fame.

Rochefoucauld says that love may be compared to a ghost since it is something we talk about but have never seen, and Lichtenberg, in his essay *Ueber die Macht der Liebe*, disputes and denies its reality and naturalness— but both are in the wrong. For if it were foreign to and contradicted human nature—in other words, if it were merely an imaginary caricature, it would not have been depicted with such zeal by the poets of all ages, or accepted by mankind with an unaltered interest; for anything artistically beautiful cannot exist without truth.

"Rien n'est beau que le vrai; le vrai seul est aimable."—BOIL.

Experience, although not that of everyday, verifies that that which as a rule begins only as a strong and yet controllable inclination, may develop, under certain conditions, into a passion, the ardour of which surpasses that of every other. It will ignore all considerations, overcome all kinds of obstacles with incredible strength and persistence. A man, in order to have his love gratified, will unhesitatingly risk his life; in fact, if his love is absolutely rejected, he will sacrifice his life into the bargain. The Werthers and Jacopo Ortis do not only exist in romances; Europe produces every year at least half-a-dozen like them: *sed ignotis perierunt mortibus illi*: for their sufferings are chronicled by the writer of official registers or by the reporters of newspapers. Indeed, readers of the police news in English and French newspapers will confirm what I have said.

Love drives a still greater number of people into the lunatic asylum. There is a case of some sort every year of two lovers committing suicide together because material circumstances happen to be unfavourable to their union. By the way, I cannot understand how it is that such people, who are confident of each other's love, and expect to find their greatest happiness in the enjoyment of it, do not avoid taking extreme steps, and prefer suffering every discomfort to sacrificing with their lives a happiness which is greater than any other they can conceive. As far as lesser phases and passages of love are concerned, all of us have them daily before our eyes, and, if we are not old, the most of us in our hearts.

After what has been brought to mind, one cannot doubt either the reality or importance of love. Instead, therefore, of wondering why a philosopher for once in a way writes on this subject, which has been constantly the theme of poets, rather should one be surprised that love, which always plays such an important *role* in a man's life, has scarcely ever been considered at all by philosophers, and that it still stands as material for them to make use of.

Plato has devoted himself more than any one else to the subject of love, especially in the *Symposium* and the *Phaedrus*; what he has said about it, however, comes within the sphere of myth, fable, and raillery, and only applies for the most part to the love of a Greek youth. The little that Rousseau says in his *Discours sur l'inegalite* is neither true nor satisfactory. Kant's disquisition on love in the third part of his treatise, *Ueber das Gefuehl*

174

des Schoenen und Erhabenen, is very superficial; it shows that he has not thoroughly gone into the subject, and therefore it is somewhat untrue. Finally, Platner's treatment of it in his *Anthropology* will be found by every one to be insipid and shallow.

To amuse the reader, on the other hand, Spinoza's definition deserves to be quoted because of its exuberant naivete: *Amor est titillatio, concomitante idea causae externae (Eth.* iv., prop. 44). It is not my intention to be either influenced or to contradict what has been written by my predecessors; the subject has forced itself upon me objectively, and has of itself become inseparable from my consideration of the world. Moreover, I shall expect least approval from those people who are for the moment enchained by this passion, and in consequence try to express their exuberant feelings in the most sublime and ethereal images. My view will seem to them too physical, too material, however metaphysical, nay, transcendent it is fundamentally.

First of all let them take into consideration that the creature whom they are idealising to-day in madrigals and sonnets would have been ignored almost entirely by them if she had been born eighteen years previously.

Every kind of love, however ethereal it may seem to be, springs entirely from the instinct of sex; indeed, it is absolutely this instinct, only in a more definite, specialised, and perhaps, strictly speaking, more individualised form. If, bearing this in mind, one considers the important *role* which love plays in all its phases and degrees, not only in dramas and novels, but also in the real world, where next to one's love of life it shows itself as the strongest and most active of all motives; if one considers that it constantly occupies half the capacities and thoughts of the younger part of humanity, and is the final goal of almost every human effort; that it influences adversely the most important affairs; that it hourly disturbs the most earnest occupations; that it sometimes deranges even the greatest intellects for a time; that it is not afraid of interrupting the transactions of statesmen or the investigations of men of learning; that it knows how to leave its love-letters and locks of hair in ministerial portfolios and philosophical manuscripts; that it knows equally well how to plan the most complicated and wicked affairs, to dissolve the most important relations, to break the strongest ties; that life, health, riches, rank, and happiness are sometimes sacrificed for its sake; that it makes the otherwise honest, perfidious, and a man who has been hitherto faithful a betrayer, and, altogether, appears as a hostile demon whose object is to overthrow, confuse, and upset everything it comes across: if all this is taken into consideration one will have reason to ask—"Why is there all this noise? Why all this crowding, blustering, anguish, and want? Why should such a trifle play so important a part and create disturbance and confusion in the well-regulated life of mankind?" But to the earnest investigator the spirit of truth gradually unfolds the answer: it is not a trifle one is dealing with; the importance of love is absolutely in keeping with the seriousness and zeal with which it is prosecuted. The ultimate aim of all love-affairs, whether they be of a tragic or comic nature, is really more important than all other aims in human life, and therefore is perfectly deserving of that profound seriousness with which it is pursued.

As a matter of fact, love determines nothing less than the *establishment of the next generation*. The existence and nature of the *dramatis personae* who come on to the scene when we have made our exit have been determined by some frivolous love-affair. As the being, the *existentia* of these future people is conditioned by our instinct of sex in general, so is the nature, the *essentia*, of these same people conditioned by the selection that the individual makes for his satisfaction, that is to say, by love, and is thereby in every respect irrevocably established. This is the key of the problem. In applying it, we shall understand it more fully if we analyse the various degrees of love, from the most fleeting sensation to the most ardent passion; we shall then see that the difference arises from the degree of individualisation of the choice. All the love-affairs of the present generation taken altogether are accordingly the *meditatio compositionis generationis futurae, e qua iterum pendent innumerae generationes* of mankind. Love

176

is of such high import, because it has nothing to do with the weal or woe of the present individual, as every other matter has; it has to secure the existence and special nature of the human race in future times; hence the will of the individual appears in a higher aspect as the will of the species; and this it is that gives a pathetic and sublime import to love-affairs, and makes their raptures and troubles transcendent, emotions which poets for centuries have not tired of depicting in a variety of ways. There is no subject that can rouse the same interest as love, since it concerns both the weal and woe of the species, and is related to every other which only concerns the welfare of the individual as body to surface.

This is why it is so difficult to make a drama interesting if it possesses no love motive; on the other hand, the subject is never exhausted, although it is constantly being utilised.

What manifests itself in the individual consciousness as instinct of sex in general, without being concentrated on any particular individual, is very plainly in itself, in its generalised form, the will to live. On the other hand, that which appears as instinct of sex directed to a certain individual, is in itself the will to live as a definitely determined individual. In this case the instinct of sex very cleverly wears the mask of objective admiration, although in itself it is a subjective necessity, and is, thereby, deceptive. Nature needs these stratagems in order to accomplish her ends. The purpose of every man in love, however objective and sublime his admiration may appear to be, is to beget a being of a definite nature, and that this is so, is verified by the fact that it is not mutual love but possession that is the essential. Without possession it is no consolation to a man to know that his love is requited. In fact, many a man has shot himself on finding himself in such a position. On the other hand, take a man who is very much in love; if he cannot have his love returned he is content simply with possession. Compulsory marriages and cases of seduction corroborate this, for a man whose love is not returned frequently finds consolation in giving handsome presents to a woman, in spite of her dislike, or making other sacrifices, so that he may buy her favour.

The real aim of the whole of love's romance, although the persons concerned are unconscious of the fact, is that a particular being may come into the world; and the way and manner in which it is accomplished is a secondary consideration. However much those of lofty sentiments, and especially of those in love, may refute the gross realism of my argument, they are nevertheless in the wrong. For is not the aim of definitely determining the individualities of the next generation a much higher and nobler aim than that other, with its exuberant sensations and transcendental soap-bubbles? Among all earthly aims is there one that is either more important or greater? It alone is in keeping with that deep-rooted feeling inseparable from passionate love, with that earnestness with which it appears, and the importance which it attaches to the trifles that come within its sphere. It is only in so far as we regard *this* end as the real one that the difficulties encountered, the endless troubles and vexations endured, in order to attain the object we love, appear to be in keeping with the matter. For it is the future generation in its entire individual determination which forces itself into existence through the medium of all this strife and trouble. Indeed, the future generation itself is already stirring in the careful, definite, and apparently capricious selection for the satisfaction of the instinct of sex which we call love. That growing affection of two lovers for each other is in reality the will to live of the new being, of which they shall become the parents; indeed, in the meeting of their yearning glances the life of a new being is kindled, and manifests itself as a well-organised individuality of the future. The lovers have a longing to be really united and made one being, and to live as such for the rest of their lives; and this longing is fulfilled in the children born to them, in whom the qualities inherited from both, but combined and united in one being, are perpetuated. Contrarily, if a man and woman mutually, persistently, and decidedly dislike each other, it indicates that they could only bring into the world a badly organised, discordant, and unhappy being. Therefore much must be attached to Calderon's words, when he calls the horrible Semiramis a daughter of the air, yet introduces her as a daughter of seduction, after which follows the murder of the husband.

Finally, it is the will to live presenting itself in the whole species, which so forcibly and exclusively attracts two individuals of different sex towards each other. This will anticipates in the being, of which they shall become the parents, an objectivation of its nature corresponding to its aims. This individual will inherit the father's will and character, the mother's intellect, and the constitution of both. As a rule, however, an individual takes more after the father in shape and the mother in stature, corresponding to the law which applies to the offspring of animals . . . It is impossible to explain the individuality of each man, which is quite exceptional and peculiar to him alone; and it is just as impossible to explain the passion of two people for each other, for it is equally individual and uncommon in character; indeed, fundamentally both are one and the same. The former is *explicite* what the latter was *implicite*.

We must consider as the origin of a new individual and true *punctum saliens* of its life the moment when the parents begin to love each other— *to fancy each other*, as the English appropriately express it. And, as has been said, in the meeting of their longing glances originates the first germ of a new being, which, indeed, like all germs, is generally crushed out. This new individual is to a certain extent a new (Platonic) Idea; now, as all Ideas strive with the greatest vehemence to enter the phenomenal sphere, and to do this, ardently seize upon the matter which the law of causality distributes among them all, so this particular Idea of a human individuality struggles with the greatest eagerness and vehemence for its realisation in the phenomenal. It is precisely this vehement desire which is the passion of the future parents for one another. Love has countless degrees, and its two extremes may be indicated as [Greek: Aphroditae pandaemos] and [Greek: ourania]; nevertheless, in essentials it is the same everywhere.

According to the degree, on the other hand, it will be the more powerful the more *individualised* it is—that is to say, the more the loved individual, by virtue of all her qualities, is exclusively fit to satisfy the lover's desire and needs determined by her own individuality. If we investigate further we shall understand more clearly what this involves. All amorous feeling immediately and essentially concentrates itself on health, strength, and beauty, and consequently on youth; because the will above all wishes to exhibit the specific character of the human species as the basis of all individuality. The same applies pretty well to everyday courtship ([Greek: Aphroditae pandaemos]). With this are bound up more special requirements, which we will consider individually later on, and with which, if there is any prospect of gratification, there is an increase of passion. Intense love, however, springs from a fitness of both individualities for each other; so that the will, that is to say the father's character and the mother's intellect combined, exactly complete that individual for which the will to live in general (which exhibits itself in the whole species) has a longing—a longing proportionate to this its greatness, and therefore surpassing the measure of a mortal heart; its motives being in a like manner beyond the sphere of the individual intellect. This, then, is the soul of a really great passion. The more perfectly two individuals are fitted for each other in the various respects which we shall consider further on, the stronger will be their passion for each other. As there are not two individuals exactly alike, a particular kind of woman must perfectly correspond with a particular kind of man—always in view of the child that is to be born. Real, passionate love is as rare as the meeting of two people exactly fitted for each other. By the way, it is because there is a possibility of real passionate love in us all that we understand why poets have depicted it in their works.

Because the kernel of passionate love turns on the anticipation of the child to be born and its nature, it is quite possible for friendship, without any admixture of sexual love, to exist between two young, good-looking people of different sex, if there is perfect fitness of temperament and intellectual capacity. In fact, a certain aversion for each other may exist also. The reason of this is that a child begotten by them would physically or mentally have discordant qualities. In short, the child's existence and nature would not be in harmony with the purposes of the will to live as it presents itself in the species.

In an opposite case, where there is no fitness of disposition, character, and mental capacity, whereby aversion, nay, even enmity for each other exists, it is possible for love to spring up. Love of this kind makes them blind to everything; and if it leads to marriage it is a very unhappy one.

And now let us more thoroughly investigate the matter. Egoism is a quality so deeply rooted in every personality that it is on egotistical ends only that one may safely rely in order to rouse the individual to activity.

To be sure, the species has a prior, nearer, and greater claim on the individual than the transient individuality itself; and yet even when the individual makes some sort of conscious sacrifice for the perpetuation and future of the species, the importance of the matter will not be made sufficiently comprehensible to his intellect, which is mainly constituted to regard individual ends.

Therefore Nature attains her ends by implanting in the individual a certain illusion by which something which is in reality advantageous to the species alone seems to be advantageous to himself; consequently he serves the latter while he imagines he is serving himself. In this process he is carried away by a mere chimera, which floats before him and vanishes again immediately, and as a motive takes the place of reality. *This illusion is instinct.* In most instances instinct may be regarded as the sense of the species which presents to the will whatever is of service to the species. But because the will has here become individual it must be deceived in such a manner for it to discern by the sense of the *individual* what the sense of the species has presented to it; in other words, imagine it is pursuing ends concerning the individual, when in reality it is pursuing merely general ends (using the word general in its strictest sense).

Outward manifestation of instinct can be best observed in animals, where the part it plays is most significant; but it is in ourselves alone that we can get to know its internal process, as of everything internal.

It is true, it is thought that man has scarcely any instinct at all, or at any rate has only sufficient instinct when he is born to seek and take his mother's breast. But as a matter of fact man has a very decided, clear, and yet complicated instinct—namely, for the selection, both earnest and capricious, of another individual, to satisfy his instinct of sex. The beauty or ugliness of the other individual has nothing whatever to do with this satisfaction in itself, that is in so far as it is a matter of pleasure based upon a pressing desire of the individual. The regard, however, for this satisfaction, which is so zealously pursued, as well as the careful selection it entails, has obviously nothing to do with the chooser himself, although he fancies that it has. Its real aim is the child to be born, in whom the type of the species is to be preserved in as pure and perfect a form as possible. For instance, different phases of degeneration of the human form are the consequences of a thousand physical accidents and moral delinquencies; and yet the genuine type of the human form is, in all its parts, always restored; further, this is accomplished under the guidance of the sense of beauty, which universally directs the instinct of sex, and without which the satisfaction of the latter would deteriorate to a repulsive necessity.

Accordingly, every one in the first place will infinitely prefer and ardently desire those who are most beautiful—in other words, those in whom the character of the species is most purely defined; and in the second, every one will desire in the other individual those perfections which he himself lacks, and he will consider imperfections, which are the reverse of his own, beautiful. This is why little men prefer big women, and fair people like dark, and so on. The ecstasy with which a man is filled at the sight of a beautiful woman, making him imagine that union with her will be the greatest happiness, is simply the *sense of the species*. The preservation of the type of the species rests on this distinct preference for beauty, and this is why beauty has such power.

We will later on more fully state the considerations which this involves. It is really instinct aiming at what is best in the species which induces a man to choose a beautiful woman, although the man himself imagines that by so doing he is only seeking to increase his own pleasure. As a matter of fact, we have here an instructive solution of the secret nature of all instinct which almost always, as in this case, prompts the individual to look after the welfare of the species. The care with which an insect selects a certain flower or fruit, or piece of flesh, or the way in which the ichneumon seeks the larva of a strange insect so that it may lay its eggs in *that particular place only*, and to secure which it fears neither labour nor danger, is obviously very analogous to the care with which a man chooses a woman of a definite nature individually suited to him. He strives for her with such ardour that he frequently, in order to attain his object, will sacrifice his happiness in life, in spite of all reason, by a foolish marriage, by some love-affair which costs him his fortune, honour, and life, even by committing crimes. And all this in accordance with the will of nature which is everywhere sovereign, so that he may serve the species in the most efficient manner, although he does so at the expense of the individual.

Instinct everywhere works as with the conception of an end, and yet it is entirely without one. Nature implants instinct where the acting individual is not capable of understanding the end, or would be unwilling to pursue it. Consequently, as a rule, it is only given prominently to animals, and in particular to those of the lowest order, which have the least intelligence. But it is only in such a case as the one we are at present considering that it is also given to man, who naturally is capable of understanding the end, but would not pursue it with the necessary zeal—that is to say, he would not pursue it at the cost of his individual welfare. So that here, as in all cases of instinct, truth takes the form of illusion in order to influence the will . . .

All this, however, on its part throws light upon the instinct of animals. They, too, are undoubtedly carried away by a kind of illusion, which represents that they are working for their own pleasure, while it is for the species that they are working with such industry and self-denial. The bird builds its nest; the insect seeks a suitable place wherein to lay its eggs, or even hunts for prey, which it dislikes itself, but which must be placed beside the eggs as food for the future larvae; the bee, the wasp, and the ant apply themselves to their skilful building and extremely complex economy. All of them are undoubtedly controlled by an illusion which conceals the service of the species under the mask of an egotistical purpose.

This is probably the only way in which to make the inner or subjective process, from which spring all manifestations of instinct, intelligible to us. The outer or objective process, however, shows in animals strongly controlled by instinct, as insects for instance, a preponderance of the ganglion—*i.e., subjective* nervous system over the *objective* or cerebral system. From which it may be concluded that they are controlled not so much by objective and proper apprehension as by subjective ideas, which excite desire and arise through the influence of the ganglionic system upon the brain; accordingly they are moved by a certain illusion . . .

The great preponderance of brain in man accounts for his having fewer instincts than the lower order of animals, and for even these few easily being led astray. For instance, the sense of beauty which instinctively guides a man in his selection of a mate is misguided when it degenerates into the proneness to pederasty. Similarly, the blue-bottle *(Musca vomitoria)*, which instinctively ought to place its eggs in putrified flesh, lays them in the blossom of the *Arum dracunculus*, because it is misled by the decaying odour of this plant. That an absolutely generic instinct is the foundation of all love of sex may be confirmed by a closer analysis of the subject— an analysis which can hardly be avoided.

In the first place, a man in love is by nature inclined to be inconstant, while a woman constant. A man's love perceptibly decreases after a certain period; almost every other woman charms him more than the one he already possesses; he longs for change: while a woman's love increases from the very moment it is returned. This is because nature aims at the preservation of the species, and consequently at as great an increase in it as possible . . . This is why a man is always desiring other women, while a woman always clings to one man; for nature compels her intuitively and unconsciously to take care of the supporter and protector of the future offspring. For this reason conjugal fidelity is artificial with the man but natural to a woman. Hence a woman's infidelity, looked at objectively on account of the consequences, and subjectively on account of its unnaturalness, is much more unpardonable than a man's.

In order to be quite clear and perfectly convinced that the delight we take in the other sex, however objective it may seem to be, is nevertheless merely instinct disguised, in other words, the sense of the species striving to preserve its type, it will be necessary to investigate more closely the considerations which influence us in this, and go into details, strange as it may seem for these details to figure in a philosophical work. These considerations may be classed in the following way:—

Those that immediately concern the type of the species, *id est*, beauty; those that concern other physical qualities; and finally, those that are merely relative and spring from the necessary correction or neutralisation of the one-sided qualities and abnormities of the two individuals by each other. Let us look at these considerations separately.

The first consideration that influences our choice and feelings is *age* . . .

The second consideration is that of *health*: a severe illness may alarm us for the time being, but an illness of a chronic nature or even cachexy frightens us away, because it would be transmitted.

The third consideration is the *skeleton*, since it is the foundation of the type of the *species*. Next to old age and disease, nothing disgusts us so much as a deformed shape; even the most beautiful face cannot make amends for it—in fact, the ugliest face combined with a well-grown shape is infinitely preferable. Moreover, we are most keenly sensible of every malformation of the *skeleton*; as, for instance, a stunted, short-legged form, and the like, or a limping gait when it is not the result of some extraneous accident: while a conspicuously beautiful figure compensates for every defect. It delights us. Further, the great importance which is attached to small feet! This is because the size of the foot is an essential characteristic of the species, for no animal has the tarsus and metatarsus combined so small as man; hence the uprightness of his gait: he is a plantigrade. And Jesus Sirach has said[17] (according to the improved translation by Kraus), "A woman that is well grown and has beautiful feet is like pillars of gold in sockets of silver." The teeth, too, are important, because they are essential for nourishment, and quite peculiarly hereditary.

The fourth consideration is a certain *plumpness*, in other words, a superabundance of the vegetative function, plasticity . . . Hence excessive thinness strikingly repels us . . . The last consideration that influences us is a *beautiful face*. Here, too, the bone parts are taken into account before everything else. So that almost everything depends on a beautiful nose, while a short *retroussé* one will mar all. A slight upward or downward turn of the nose has often determined the life's happiness of a great many maidens; and justly so, for the type of the species is at stake.

A small mouth, by means of small maxillae, is very essential, as it is the specific characteristic of the human face as distinguished from the muzzle of the brutes. A receding, as it were, a cut-away chin is particularly repellent, because *mentum prominulum* is a characteristic belonging exclusively to our species.

Finally, we come to the consideration of beautiful eyes and a beautiful forehead; they depend upon the psychical qualities, and in particular, the intellectual, which are inherited from the mother. The unconscious considerations which, on the other hand, influence women in their choice naturally cannot be so accurately specified. In general, we may say the following:—That the age they prefer is from thirty to thirty-five. For instance, they prefer men of this age to youths, who in reality possess the highest form of human beauty. The reason for this is that they are not guided by taste but by instinct, which recognises in this particular age the acme of generative power. In general, women pay little attention to beauty, that is, to beauty of face; they seem to take it upon themselves alone to endow the child with beauty. It is chiefly the strength of a man and the courage that goes with it that attract them, for both of these promise the generation of robust children and at the same time a brave protector for them. Every physical defect in a man, any deviation from the type, a woman may, with regard to the child, eradicate if she is faultless in these parts herself or excels in a contrary direction. The only exceptions are those qualities which are peculiar to the man, and which, in consequence, a mother cannot bestow on her child; these include the masculine build of the skeleton, breadth of shoulder, small hips, straight legs, strength of muscle, courage, beard, and so on. And so it happens that a woman frequently loves an ugly man, albeit she never loves an unmanly man, because she cannot neutralise his defects.

The second class of considerations that are the source of love are those depending on the psychical qualities. Here we shall find that a woman universally is attracted by the qualities of a man's heart or character, both of which are inherited from the father. It is mainly firmness of will, determination and courage, and may be honesty and goodness of heart too, that win a woman over; while intellectual qualifications exercise no direct or instinctive power over her, for the simple reason that these are not inherited from the father. A lack of intelligence carries no weight with her; in fact, a superabundance of mental power or even genius, as abnormities, might have an unfavourable effect. And so we frequently find a woman preferring a stupid, ugly, and ill-mannered man to one who is well-educated, intellectual, and agreeable. Hence, people of extremely different temperament frequently marry for love—that is to say, *he* is coarse, strong, and narrow-minded, while *she* is very sensitive, refined, cultured, and aesthetic, and so on; or *he* is genial and clever, and *she* is a goose.

> "Sic visum Veneri; cui placet impares
> Formas atque animos sub juga aenea
>   Saevo mittere cum joco."

187

The reason for this is, that she is not influenced by intellectual considerations, but by something entirely different, namely, instinct. Marriage is not regarded as a means for intellectual entertainment, but for the generation of children; it is a union of hearts and not of minds. When a woman says that she has fallen in love with a man's mind, it is either a vain and ridiculous pretence on her part or the exaggeration of a degenerate being. A man, on the other hand, is not controlled in instinctive love by the *qualities* of the woman's *character*; this is why so many a Socrates has found his Xantippe, as for instance, Shakespeare, Albrecht Duerer, Byron, and others. But here we have the influence of intellectual qualities, because they are inherited from the mother; nevertheless their influence is easily overpowered by physical beauty, which concerns more essential points, and therefore has a more direct effect. By the way, it is for this reason that mothers who have either felt or experienced the former influence have their daughters taught the fine arts, languages, etc., so that they may prove more attractive. In this way they hope by artificial means to pad the intellect, just as they do their bust and hips if it is necessary to do so. Let it be understood that here we are simply speaking of that attraction which is absolutely direct and instinctive, and from which springs real love. That an intelligent and educated woman esteems intelligence and brains in a man, and that a man after deliberate reasoning criticises and considers the character of his *fiancee*, are matters which do not concern our present subject. Such things influence a rational selection in marriage, but they do not control passionate love, which is our matter.

Up to the present I have taken into consideration merely the *absolute* considerations—*id est*, such considerations as apply to every one. I now come to the *relative* considerations, which are individual, because they aim at rectifying the type of the species which is defectively presented and at correcting any deviation from it existing in the person of the chooser himself, and in this way lead back to a pure presentation of the type. Hence each man loves what he himself is deficient in. The choice that is based on relative considerations—that is, has in view the constitution of the individual—is much more certain, decided, and exclusive than the choice that is made after merely absolute considerations; consequently real passionate love will have its origin, as a rule, in these relative considerations, and it will only be the ordinary phases of love that spring from the absolute. So that it is not stereotyped, perfectly beautiful women who are wont to kindle great passions. Before a truly passionate feeling can exist, something is necessary that is perhaps best expressed by a metaphor in chemistry—namely, the two persons must neutralise each other, like acid and alkali to a neutral salt. Before this can be done the following conditions are essential. In the first place, all sexuality is one-sided. This one-sidedness is more definitely expressed and exists in a higher degree in one person than in another; so that it may be better supplemented and neutralised in each individual by one person than by another of the opposite sex, because the individual requires a one-sidedness opposite to his own in order to complete the type of humanity in the new individual to be generated, to the constitution of which everything tends . . .

The following is necessary for this neutralisation of which we are speaking. The particular degree of *his* manhood must exactly correspond to the particular degree of *her* womanhood in order to exactly balance the one-sidedness of each. Hence the most manly man will desire the most womanly woman, and *vice versa*, and so each will want the individual that exactly corresponds to him in degree of sex. Inasmuch as two persons fulfil this necessary relation towards each other, it is instinctively felt by them and is the origin, together with the other *relative* considerations, of the higher degrees of love. While, therefore, two lovers are pathetically talking about the harmony of their souls, the kernel of the conversation is for the most part the harmony concerning the individual and its perfection, which obviously is of much more importance than the harmony of their souls—which frequently turns out to be a violent discord shortly after marriage.

We now come to those other relative considerations which depend on each individual trying to eradicate, through the medium of another, his weaknesses, deficiencies, and deviations from the type, in order that they may not be perpetuated in the child that is to be born or develop into absolute abnormities. The weaker a man is in muscular power, the more will he desire a woman who is muscular; and the same thing applies to a woman . . .

Nevertheless, if a big woman choose a big husband, in order, perhaps, to present a better appearance in society, the children, as a rule, suffer for her folly. Again, another very decided consideration is complexion. Blonde people fancy either absolutely dark complexions or brown; but it is rarely the case *vice versa*. The reason for it is this: that fair hair and blue eyes are a deviation from the type and almost constitute an abnormity, analogous to white mice, or at any rate white horses. They are not indigenous to any other part of the world but Europe,—not even to the polar regions,—and are obviously of Scandinavian origin. *En passant*, it is my conviction that a white skin is not natural to man, and that by nature he has either a black or brown skin like our forefathers, the Hindoos, and that the white man was never originally created by nature; and that, therefore, there is no *race* of white people, much as it is talked about, but every white man is a bleached one. Driven up into the north, where he was a stranger, and where he existed only like an exotic plant, in need of a hothouse in winter, man in the course of centuries became white. The gipsies, an Indian tribe which emigrated only about four centuries ago, show the transition of the Hindoo's complexion to ours. In love, therefore, nature strives to return to dark hair and brown eyes, because they are the original type; still, a white skin has become second nature, although not to such an extent as to make the dark skin of the Hindoo repellent to us.

Finally, every man tries to find the corrective of his own defects and aberrations in the particular parts of his body, and the more conspicuous the defect is the greater is his determination to correct it. This is why snub-nosed persons find an aquiline nose or a parrot-like face so indescribably pleasing; and the same thing applies to every other part of the body. Men of immoderately long and attenuated build delight in a stunted and short figure. Considerations of temperament also influence a man's choice. Each prefers a temperament the reverse of his own; but only in so far as his is a decided one.

A man who is quite perfect in some respect himself does not, it is true, desire and love imperfection in this particular respect, yet he can be more easily reconciled to it than another man, because he himself saves the children from being very imperfect in this particular. For instance, a man who has a very white skin himself will not dislike a yellowish complexion, while a man who has a yellowish complexion will consider a dazzlingly white skin divinely beautiful. It is rare for a man to fall in love with a positively ugly woman, but when he does, it is because exact harmony in the degree of sex exists between them, and all her abnormities are precisely the opposite to, that is to say, the corrective of his. Love in these circumstances is wont to attain a high degree.

The profoundly earnest way in which we criticise and narrowly consider every part of a woman, while she on her part considers us; the scrupulously careful way we scrutinise, a woman who is beginning to please us; the fickleness of our choice; the strained attention with which a man watches his *fiancee*; the care he takes not to be deceived in any trait; and the great importance he attaches to every more or less essential trait,—all this is quite in keeping with the importance of the end. For the child that is to be born will have to bear a similar trait through its whole life; for instance, if a woman stoops but a little, it is possible for her son to be inflicted with a hunchback; and so in every other respect. We are not conscious of all this, naturally. On the contrary, each man imagines that his choice is made in the interest of his own pleasure (which, in reality, cannot be interested in it at all); his choice, which we must take for granted is in keeping with his own individuality, is made precisely in the interest of the species, to maintain the type of which as pure

191

as possible is the secret task. In this case the individual unconsciously acts in the interest of something higher, that is, the species. This is why he attaches so much importance to things to which he might, nay, would be otherwise indifferent. There is something quite singular in the unconsciously serious and critical way two young people of different sex look at each other on meeting for the first time; in the scrutinising and penetrating glances they exchange, in the careful inspection which their various traits undergo. This scrutiny and analysis represent the *meditation of the genius of the species* on the individual which may be born and the combination of its qualities; and the greatness of their delight in and longing for each other is determined by this meditation. This longing, although it may have become intense, may possibly disappear again if something previously unobserved comes to light. And so the genius of the species meditates concerning the coming race in all who are yet not too old. It is Cupid's work to fashion this race, and he is always busy, always speculating, always meditating. The affairs of the individual in their whole ephemeral totality are very trivial compared with those of this divinity, which concern the species and the coming race; therefore he is always ready to sacrifice the individual regardlessly. He is related to these ephemeral affairs as an immortal being is to a mortal, and his interests to theirs as infinite to finite. Conscious, therefore, of administering affairs of a higher order than those that concern merely the weal and woe of the individual, he administers them with sublime indifference amid the tumult of war, the bustle of business, or the raging of a plague—indeed, he pursues them into the seclusion of the cloisters.

It has been seen that the intensity of love grows with its individuation; we have shown that two individuals may be so physically constituted, that, in order to restore the best possible type of the species, the one is the special and perfect complement of the other, which, in consequence, exclusively desires it. In a case of this kind, passionate love arises, and as it is bestowed on one object, and one only—that is to say, because it appears in the *special* service of the species—it immediately assumes

192

a nobler and sublimer nature. On the other hand, mere sexual instinct is base, because, without individuation, it is directed to all, and strives to preserve the species merely as regards quantity with little regard for quality. Intense love concentrated on one individual may develop to such a degree, that unless it is gratified all the good things of this world, and even life itself, lose their importance. It then becomes a desire, the intensity of which is like none other; consequently it will make any kind of sacrifice, and should it happen that it cannot be gratified, it may lead to madness or even suicide. Besides these unconscious considerations which are the source of passionate love, there must be still others, which we have not so directly before us. Therefore, we must take it for granted that here there is not only a fitness of constitution but also a special fitness between the man's *will* and the woman's *intellect*, in consequence of which a perfectly definite individual can be born to them alone, whose existence is contemplated by the genius of the species for reasons to us impenetrable, since they are the very essence of the thing-in-itself. Or more strictly speaking, the will to live desires to objectivise itself in an individual which is precisely determined, and can only be begotten by this particular father and this particular mother. This metaphysical yearning of the will in itself has immediately, as its sphere of action in the circle of human beings, the hearts of the future parents, who accordingly are seized with this desire. They now fancy that it is for their own sakes they are longing for what at present has purely a metaphysical end, that is to say, for what does not come within the range of things that exist in reality. In other words, it is the desire of the future individual to enter existence, which has first become possible here, a longing which proceeds from the primary source of all being and exhibits itself in the phenomenal world as the intense love of the future parents for each other, and has little regard for anything outside itself. In fact, love is an illusion like no other; it will induce a man to sacrifice everything he possesses in the world, in order to obtain this woman, who in reality will satisfy him no more than any other. It also ceases to exist when the end, which was in reality metaphysical, has been frustrated perhaps by the woman's barrenness (which, according to Hufeland, is the result of nineteen accidental defects in the constitution), just as it is frustrated daily in millions of crushed germs in which the same metaphysical life-principle struggles to exist; there is no other consolation in this than that there is an infinity of space, time, and matter, and consequently inexhaustible opportunity, at the service of the will to live.

Although this subject has not been treated by Theophrastus Paracelsus, and my entire train of thought is foreign to him, yet it must have presented itself to him, if even in a cursory way, when he gave utterance to the following remarkable words, written in quite a different context and in his usual desultory style: *Hi sunt, quos Deus copulavit, ut eam, quae fuit Uriae et David; quamvis ex diametro (sic enim sibi humana mens persuadebat) cum justo et legitimo matrimonio pugnaret hoc . . . sed propter Salomonem, qui aliunde nasci non potuit, nisi ex Bathseba, conjuncto David semine, quamvis meretrice, conjunxit eos Deus.*[18]

The yearning of love, the [Greek: himeros], which has been expressed in countless ways and forms by the poets of all ages, without their exhausting the subject or even doing it justice; this longing which makes us imagine that the possession of a certain woman will bring interminable happiness, and the loss of her, unspeakable pain; this longing and this pain do not arise from the needs of an ephemeral individual, but are, on the contrary, the sigh of the spirit of the species, discerning irreparable means of either gaining or losing its ends. It is the species alone that has an interminable existence: hence it is capable of endless desire, endless gratification, and endless pain. These, however, are imprisoned in the heart of a mortal; no wonder, therefore, if it seems like to burst, and can find no expression for the announcements of endless joy or endless pain. This it is that forms the substance of all erotic poetry that is sublime in character, which, consequently, soars into transcendent metaphors, surpassing everything earthly. This is the theme of Petrarch, the material for the St. Preuxs, Werthers, and Jacopo Ortis, who otherwise could be neither understood nor explained. This infinite regard is not based on any kind of intellectual, nor, in general, upon any real merits of the beloved one; because the lover frequently does not know her well enough; as was the case with Petrarch.

It is the spirit of the species alone that can see at a glance of what *value* the beloved one is to *it* for its purposes. Moreover, great passions, as a rule, originate at first sight:

> "Who ever lov'd, that lov'd not at first sight."
> —SHAKESPEARE, *As You Like It,* iii. 5.

Curiously enough, there is a passage touching upon this in *Guzmann de Alfarache,* a well-known romance written two hundred and fifty years ago by Mateo Aleman: *No es necessario para que uno ame, que pase distancia de tiempo, que siga discurso, in haga eleccion, sino que con aquella primera y sola vista, concurran juntamente cierta correspondencia o consonancia, o lo que aca solemos vulgarmente decir, una confrontacion de sangre, a que por particular influxo suelen mover las estrellas.* (For a man to love there is no need for any length of time to pass for him to weigh considerations or make his choice, but only that a certain correspondence and consonance is encountered on both sides at the first and only glance, or that which is ordinarily called *a sympathy of blood,* to which a peculiar influence of the stars generally impels.) Accordingly, the loss of the beloved one through a rival, or through death, is the greatest pain of all to those passionately in love; just because it is of a transcendental nature, since it affects him not merely as an individual, but also assails him in his *essentia aeterna,* in the life of the species, in whose special will and service he was here called. This is why jealousy is so tormenting and bitter, and the giving up of the loved one the greatest of all sacrifices. A hero is ashamed of showing any kind of emotion but that which may be the outcome of love; the reason for this is, that when he is in love it is not he, but the species which is grieving. In Calderon's *Zenobia the Great* there is a scene in the second act between Zenobia and Decius where the latter says, *Cielos, luego tu me quieres? Perdiera cien mil victorias, Volvierame,* etc. (Heavens! then you love me? For this I would sacrifice a thousand victories, etc.) In this case honour, which has hitherto outweighed every other interest, is driven out of the field directly love—*i.e.,* the interest of the species—comes into play and discerns something that will be of decided advantage to itself; for the interest of the species, compared with that of the mere individual, however important this may be, is infinitely more important. Honour, duty, and fidelity succumb to it after they have withstood every other temptation—the menace of death even. We find the same going on in private life; for instance, a man has less conscience when in love than in any other circumstances. Conscience is sometimes put on one side even by people who are otherwise honest and straightforward, and infidelity recklessly committed if they are passionately in love—*i.e.,* when the interest of the species has taken possession of them. It would seem, indeed, as if they believed themselves conscious of a greater authority than the interests of individuals could ever confer; this is simply because they are concerned in the interest of the species. Chamfort's utterance in this respect is remarkable: *Quand un homme et une femme ont l'un pour l'autre une passion violente, il me semble toujours que quelque soient les obstacles qui les separent, un mari, des parens, etc.; les deux amans sont l'un a l'autre, de par la Nature, qu'ils s'appartiennent de droit devin, malgre les lois et les conventions humaines . . .* From this standpoint the greater part of the *Decameron* seems a mere mocking and jeering on the part of the genius of the species at the rights and interests of the individual which it treads underfoot. Inequality of rank and all similar relations are put on one side with the same indifference and disregarded by the genius of the species, if they thwart the union of two people passionately in love with one another: it pursues its ends pertaining to endless generations, scattering human principles and scruples abroad like chaff.

For the same reason, a man will willingly risk every kind of danger, and even become courageous, although he may otherwise be faint-hearted.

196

What a delight we take in watching, either in a play or novel, two young lovers fighting for each other—i.e., for the interest of the species—and their defeat of the old people, who had only in view the welfare of the individual! For the struggling of a pair of lovers seems to us so much more important, delightful, and consequently justifiable than any other, as the species is more important than the individual.

Accordingly, we have as the fundamental subject of almost all comedies the genius of the species with its purposes, running counter to the personal interests of the individuals presented, and, in consequence, threatening to undermine their happiness. As a rule it carries out its ends, which, in keeping with true poetic justice, satisfies the spectator, because the latter feels that the purposes of the species widely surpass those of the individual. Hence he is quite consoled when he finally takes leave of the victorious lovers, sharing with them the illusion that they have established their own happiness, while, in truth, they have sacrificed it for the welfare of the species, in opposition to the will of the discreet old people.

It has been attempted in a few out-of-the-way comedies to reverse this state of things and to effect the happiness of the individuals at the cost of the ends of the species; but here the spectator is sensible of the pain inflicted on the genius of the species, and does not find consolation in the advantages that are assured to the individuals.

Two very well-known little pieces occur to me as examples of this kind: *La reine de 16 ans*, and *Le mariage de raison*.

In the love-affairs that are treated in tragedies the lovers, as a rule, perish together: the reason for this is that the purposes of the species, whose tools the lovers were, have been frustrated, as, for instance, in *Romeo and Juliet*, *Tancred*, *Don Carlos*, *Wallenstein*, *The Bride of Messina*, and so on.

A man in love frequently furnishes comic as well as tragic aspects; for being in the possession of the spirit of the species and controlled by it, he no longer belongs to himself, and consequently his line of conduct is not in keeping with that of the individual. It is fundamentally this that in the higher phases of love gives such a poetical and sublime colour, nay, transcendental and hyperphysical turn to a man's thoughts, whereby he appears to lose sight of his essentially material purpose. He is inspired by the spirit of the species, whose affairs are infinitely more important than any which concern mere individuals, in order to establish by special mandate of this spirit the existence of an indefinitely long posterity with *this* particular and precisely determined nature, which it can receive only from him as father and his loved one as mother, and which, moreover, *as such* never comes into existence, while the objectivation of the will to live expressly demands this existence. It is the feeling that he is engaged in affairs of such transcendent importance that exalts the lover above everything earthly, nay, indeed, above himself, and gives such a hyperphysical clothing to his physical wishes, that love becomes, even in the life of the most prosaic, a poetical episode; and then the affair often assumes a comical aspect. That mandate of the will which objectifies itself in the species presents itself in the consciousness of the lover under the mask of the anticipation of an infinite happiness, which is to be found in his union with this particular woman. This illusion to a man deeply in love becomes so dazzling that if it cannot be attained, life itself not only loses all charm, but appears to be so joyless, hollow, and uninteresting as to make him too disgusted with it to be afraid of the terrors of death; this is why he sometimes of his own free will cuts his life short. The will of a man of this kind has become engulfed in that of the species, or the will of the species has obtained so great an ascendency over the will of the individual that if such a man cannot be effective in the manifestation of the first, he disdains to be so in the last. The individual in this case is too weak a vessel to bear the infinite longing of the will of the species concentrated upon a definite object. When this is the case suicide is the result, and sometimes suicide of the two lovers; unless nature, to prevent this, causes insanity, which then enshrouds with its veil the consciousness of so hopeless a condition. The truth of this is confirmed yearly by various cases of this description.

However, it is not only unrequited love that leads frequently to a tragic end; for requited love more frequently leads to unhappiness than to happiness. This is because its demands often so severely clash with the personal welfare of the lover concerned as to undermine it, since the demands are incompatible with the lover's other circumstances, and in consequence destroy the plans of life built upon them. Further, love frequently runs counter not only to external circumstances but to the individuality itself, for it may fling itself upon a person who, apart from the relation of sex, may become hateful, despicable, nay, even repulsive. As the will of the species, however, is so very much stronger than that of the individual, the lover shuts his eyes to all objectionable qualities, overlooks everything, ignores all, and unites himself for ever to the object of his passion. He is so completely blinded by this illusion that as soon as the will of the species is accomplished the illusion vanishes and leaves in its place a hateful companion for life. From this it is obvious why we often see very intelligent, nay, distinguished men married to dragons and she-devils, and why we cannot understand how it was possible for them to make such a choice. Accordingly, the ancients represented *Amor* as blind. In fact, it is possible for a lover to clearly recognise and be bitterly conscious of horrid defects in his *fiancee's* disposition and character—defects which promise him a life of misery—and yet for him not to be filled with fear:

> "I ask not, I care not,
>   If guilt's in thy heart;
> I know that I love thee,
>   Whatever thou art."

For, in truth, he is not acting in his own interest but in that of a third person, who has yet to come into existence, albeit he is under the

199

impression that he is acting in his own But it is this very *acting in some one else's interest* which is everywhere the stamp of greatness and gives to passionate love the touch of the sublime, making it a worthy subject for the poet. Finally, a man may both love and hate his beloved at the same time. Accordingly, Plato compares a man's love to the love of a wolf for a sheep. We have an instance of this kind when a passionate lover, in spite of all his exertions and entreaties, cannot obtain a hearing upon any terms.

> "I love and hate her."
> —SHAKESPEARE, *Cymb.* iii. 5.

When hatred is kindled, a man will sometimes go so far as to first kill his beloved and then himself. Examples of this kind are brought before our notice yearly in the newspapers. Therefore Goethe says truly:

> "Bei aller verschmaehten Liebe, beim hoellichen Elemente!
> Ich wollt', ich wuesst' was aerger's, das ich fluchen koennte!"

It is in truth no hyperbole on the part of a lover when he calls his beloved's coldness, or the joy of her vanity, which delights in his suffering, *cruelty*. For he has come under the influence of an impulse which, akin to the instinct of animals, compels him in spite of all reason to unconditionally pursue his end and discard every other; he cannot give it up. There has not been one but many a Petrarch, who, failing to have his love requited, has been obliged to drag through life as if his feet were either fettered or carried a leaden weight, and give vent to his sighs in a lonely forest; nevertheless there was only one Petrarch who possessed the true poetic instinct, so that Goethe's beautiful lines are true of him:

> "Und wenn der Mensch in seiner Quaal verstummt,
> Gab mir ein Gott, zu sagen, wie ich leide."

200

As a matter of fact, the genius of the species is at continual warfare with the guardian genius of individuals; it is its pursuer and enemy; it is always ready to relentlessly destroy personal happiness in order to carry out its ends; indeed, the welfare of whole nations has sometimes been sacrificed to its caprice. Shakespeare furnishes us with such an example in *Henry VI* Part III., Act iii., Scenes 2 and 3. This is because the species, in which lies the germ of our being, has a nearer and prior claim upon us than the individual, so that the affairs of the species are more important than those of the individual. Sensible of this, the ancients personified the genius of the species in Cupid, notwithstanding his having the form of a child, as a hostile and cruel god, and therefore one to be decried as a capricious and despotic demon, and yet lord of both gods and men.

[Greek: Su d' o theon tyranne k' anthropon, Eros.]
(Tu, deorum hominumque tyranne, Amor!)

Murderous darts, blindness, and wings are Cupid's attributes. The latter signify inconstancy, which as a rule comes with the disillusion following possession.

Because, for instance, love is based on an illusion and represents what is an advantage to the species as an advantage to the individual, the illusion necessarily vanishes directly the end of the species has been attained. The spirit of the species, which for the time being has got the individual into its possession, now frees him again. Deserted by the spirit, he relapses into his original state of narrowness and want; he is surprised to find that after all his lofty, heroic, and endless attempts to further his own pleasure he has obtained but little; and contrary to his expectation, he finds that he is no happier than he was before. He discovers that he has been the dupe of the will of the species. Therefore, as a rule, a Theseus who has been made happy will desert his Ariadne. If Petrarch's passion had been gratified his song would have become silent from that moment, as that of the birds as soon as the eggs are laid.

Let it be said in passing that, however much my metaphysics of love may displease those in love, the fundamental truth revealed by me would enable them more effectually than anything else to overcome their passion, if considerations of reason in general could be of any avail. The words of the comic poet of ancient times remain good: *Quae res in se neque consilium, neque modum habet ullum, eam consilio regere non potes.* People who marry for love do so in the interest of the species and not of the individuals. It is true that the persons concerned imagine they are promoting their own happiness; but their real aim, which is one they are unconscious of, is to bring forth an individual which can be begotten by them alone. This purpose having brought them together, they ought henceforth to try and make the best of things. But it very frequently happens that two people who have been brought together by this instinctive illusion, which is the essence of passionate love, are in every other respect temperamentally different. This becomes apparent when the illusion wears off, as it necessarily must.

Accordingly, people who marry for love are generally unhappy, for such people look after the welfare of the future generation at the expense of the present. *Quien se casa por amores, ha de vivir con dolores* (He who marries for love must live in grief), says the Spanish proverb. Marriages *de convenance*, which are generally arranged by the parents, will turn out the reverse. The considerations in this case which control them, whatever their nature may be, are at any rate real and unable to vanish of themselves. A marriage of this kind attends to the welfare of the present generation to the detriment of the future, it is true; and yet this remains problematical.

A man who marries for money, and not for love, lives more in the interest of the individual than in that of the species; a condition exactly opposed to truth; therefore it is unnatural and rouses a certain feeling of contempt. A girl who against the wish of her parents refuses to marry a rich man, still young, and ignores all considerations of *convenance*, in order to choose another instinctively to her liking, sacrifices her individual welfare to the species. But it is for this very reason that she meets with a certain approval, for she has given preference to what was more important and acted in the spirit of nature (of the species) more exactly; while the parents advised only in the spirit of individual egoism.

As the outcome of all this, it seems that to marry means that either the interest of the individual or the interest of the species must suffer. As a rule one or the other is the case, for it is only by the rarest and luckiest accident that *convenance* and passionate love go hand in hand. The wretched condition of most persons physically, morally, and intellectually may be partly accounted for by the fact that marriages are not generally the result of pure choice and inclination, but of all kinds of external considerations and accidental circumstances. However, if inclination to a certain degree is taken into consideration, as well as convenience, this is as it were a compromise with the genius of the species. As is well known, happy marriages are few and far between, since marriage is intended to have the welfare of the future generation at heart and not the present.

However, let me add for the consolation of the more tender-hearted that passionate love is sometimes associated with a feeling of quite another kind—namely, real friendship founded on harmony of sentiment, but this, however, does not exist until the instinct of sex has been extinguished. This friendship will generally spring from the fact that the physical, moral, and intellectual qualities which correspond to and supplement each other in two individuals in love, in respect of the child to be born, will also supplement each other in respect of the individuals themselves as opposite qualities of temperament and intellectual excellence, and thereby establish a harmony of sentiment.

The whole metaphysics of love which has been treated here is closely related to my metaphysics in general, and the light it throws upon this may be said to be as follows.

We have seen that a man's careful choice, developing through innumerable degrees to passionate love, for the satisfaction of his instinct of sex, is based upon the fundamental interest he takes in the constitution of the next generation. This overwhelming interest that he takes verifies two truths which have been already demonstrated.

First: Man's immortality, which is perpetuated in the future race. For this interest of so active and zealous a nature, which is neither the result of reflection nor intention, springs from the innermost characteristics and tendencies of our being, could not exist so continuously or exercise such great power over man if the latter were really transitory and if a race really and totally different to himself succeeded him merely in point of time.

Second: That his real nature is more closely allied to the species than to the individual. For this interest that he takes in the special nature of the species, which is the source of all love, from the most fleeting emotion to the most serious passion, is in reality the most important affair in each man's life, the successful or unsuccessful issue of which touches him more nearly than anything else. This is why it has been pre-eminently called the "affair of the heart." Everything that merely concerns one's own person is set aside and sacrificed, if the case require it, to this interest when it is of a strong and decided nature. Therefore in this way man proves that he is more interested in the species than in the individual, and that he lives more directly in the interest of the species than in that of the individual.

Why, then, is a lover so absolutely devoted to every look and turn of his beloved, and ready to make any kind of sacrifice for her? Because the *immortal* part of him is yearning for her; it is only the *mortal* part of him that longs for everything else. That keen and even intense longing for a particular woman is accordingly a direct pledge of the immortality of the essence of our being and of its perpetuity in the species.

To regard this perpetuity as something unimportant and insufficient is an error, arising from the fact that in thinking of the continuity of the species we only think of the future existence of beings similar to ourselves, but in no respect, however, identical with us; and again, starting from knowledge directed towards without, we only grasp the outer form of the species as it presents itself to us, and do not take into consideration its inner nature. It is precisely this inner nature that lies at the foundation of our own consciousness as its kernel, and therefore is more direct than our consciousness itself, and as thing-in-itself exempt from the *principium individuationis*—is in reality identical and the same in all individuals, whether they exist at the same or at different times.

This, then, is the will to live—that is to say, it is exactly *that which* so intensely desires both life and continuance, and which accordingly remains unharmed and unaffected by death. Further, its present state cannot be improved, and while there is life it is certain of the unceasing sufferings and death of the individual. The *denial* of the will to live is reserved to free it from this, as the means by which the individual will breaks away from the stem of the species, and surrenders that existence in it.

We are wanting both in ideas and all data as to what it is after that. We can only indicate it as something which is free to be will to live or not to live. Buddhism distinguishes the latter case by the word *Nirvana*. It is the point which as such remains for ever impenetrable to all human knowledge.

Looking at the turmoil of life from this standpoint we find all occupied with its want and misery, exerting all their strength in order to satisfy its endless needs and avert manifold suffering, without, however, daring to expect anything else in return than merely the preservation of this tormented individual existence for a short span of time. And yet, amid all this turmoil we see a pair of lovers exchanging longing glances—yet why so secretly, timidly, and stealthily? Because these lovers are traitors secretly striving to perpetuate all this misery and turmoil that otherwise would come to a timely end.

FOOTNOTES:

17   Ch. xxvi. 23.

18   *De vita longa* i. 5.

PHYSIOGNOMY.

That the outside reflects the inner man, and that the face expresses his whole character, is an obvious supposition and accordingly a safe one, demonstrated as it is in the desire people have *to see* on all occasions a man who has distinguished himself by something good or evil, or produced some exceptional work; or if this is denied them, at any rate to hear from others *what he looks like*. This is why, on the one hand, they go to places where they conjecture he is to be found; and on the other, why the press, and especially the English press, tries to describe him in a minute and striking way; he is soon brought visibly before us either by a painter or an engraver; and finally, photography, on that account so highly prized, meets this necessity in a most perfect way.

It is also proved in everyday life that each one inspects the physiognomy of those he comes in contact with, and first of all secretly tries to discover their moral and intellectual character from their features. This could not be the case if, as some foolish people state, the outward appearance of a man is of no importance; nay, if the soul is one thing and the body another, and the latter related to the soul as the coat is to the man himself.

Rather is every human face a hieroglyph, which, to be sure, admits of being deciphered—nay, the whole alphabet of which we carry about with us. Indeed, the face of a man, as a rule, bespeaks more interesting matter than his tongue, for it is the compendium of all which he will ever say, as it is the register of all his thoughts and aspirations. Moreover, the tongue only speaks the thoughts of one man, while the face expresses a thought of nature. Therefore it is worth while to observe everybody attentively; even if they are not worth talking to. Every individual is worthy of observation as a single thought of nature; so is beauty in the highest degree, for it is a higher and more general conception of nature: it is her thought of a species. This is why we are so captivated by beauty. It is a fundamental and principal thought of Nature; whereas the individual is only a secondary thought, a corollary.

In secret, everybody goes upon the principle that a man *is* what he *looks*; but the difficulty lies in its application. The ability to apply it is partly innate and partly acquired by experience; but no one understands it thoroughly, for even the most experienced may make a mistake. Still, it is not the face that deceives, whatever Figaro may say, but it is we who are deceived in reading what is not there. The deciphering of the face is certainly a great and difficult art. Its principles can never be learnt *in abstracto*. Its first condition is that the man must be looked at from a *purely objective* point of view; which is not so easy to do. As soon as, for instance, there is the slightest sign of dislike, or affection, or fear, or hope, or even the thought of the impression which we ourselves are making on him—in short, as soon as anything of a subjective nature is present, the hieroglyphics become confused and falsified. The sound of a language is only heard by one who does not understand it, because in thinking of the significance one is not conscious of the sign itself; and similarly the physiognomy of a man is only seen by one to whom it is still strange— that is to say, by one who has not become accustomed to his face through seeing him often or talking to him. Accordingly it is, strictly speaking, the first glance that gives one a purely objective impression of a face, and makes it possible for one to decipher it. A smell only affects us when we first perceive it, and it is the first glass of wine which gives us its real taste; in the same way, it is only when we see a face for the first time that it makes a full impression upon us. Therefore one should carefully attend to the first impression; one should make a note of it, nay, write it down if the man is of personal importance—that is, if one can trust one's own sense of physiognomy. Subsequent acquaintance and intercourse will erase that impression, but it will be verified one day in the future.

En passant, let us not conceal from ourselves the fact that this first impression is as a rule extremely disagreeable: but how little there is in the majority of faces! With the exception of those that are beautiful, good-natured, and intellectual—that is, the very few and exceptional,—I believe a new face for the most part gives a sensitive person a sensation akin to a shock, since the disagreeable impression is presented in a new and surprising combination.

As a rule it is indeed *a sorry sight*. There are individuals whose faces are stamped with such naive vulgarity and lowness of character, such an animal limitation of intelligence, that one wonders how they care to go out with such a face and do not prefer to wear a mask. Nay, there are faces a mere glance at which makes one feel contaminated. One cannot therefore blame people, who are in a position to do so, if they seek solitude and escape the painful sensation of *"seeing new faces."* The *metaphysical* explanation of this rests on the consideration that the individuality of each person is exactly that by which he should be reclaimed and corrected.

If any one, on the other hand, will be content with a *psychological* explanation, let him ask himself what kind of physiognomy can be expected in those whose minds, their whole life long, have scarcely ever entertained anything but petty, mean, and miserable thoughts, and vulgar, selfish, jealous, wicked, and spiteful desires. Each one of these thoughts and desires has left its impress on the face for the length of time it existed; all these marks, by frequent repetition, have eventually become furrows and blemishes, if one may say so. Therefore the appearance of the majority of people is calculated to give one a shock at first sight, and it is only by degrees that one becomes accustomed to a face—that is to say, becomes so indifferent to the impression as to be no longer affected by it.

But that the predominating facial expression is formed by countless fleeting and characteristic contortions is also the reason why the faces of intellectual men only become moulded gradually, and indeed only attain their sublime expression in old age; whilst portraits of them in their youth only show the first traces of it. But, on the other hand, what has just been said about the shock one receives at first sight coincides with the above remark, that it is only at first sight that a face makes its true and full impression. In order to get a purely objective and true impression of it, we must stand in no kind of relation to the person, nay, if possible, we must not even have spoken to him. Conversation makes one in some measure friendly disposed, and brings us into a certain *rapport*, a reciprocal *subjective* relation, which immediately interferes with our taking an objective view. As everybody strives to win either respect or friendship for himself, a man who is being observed will immediately resort to every art of dissembling, and corrupt us with his airs, hypocrisies, and flatteries; so that in a short time we no longer see what the first impression had clearly shown us. It is said that "most people gain on further acquaintance" but what ought to be said is that "they delude us" on further acquaintance. But when these bad traits have an opportunity of showing themselves later on, our first impression generally receives its justification. Sometimes a further acquaintance is a hostile one, in which case it will not be found that people gain by it. Another reason for the apparent advantage of a further acquaintance is, that the man whose first appearance repels us, as soon as we converse with him no longer shows his true being and character, but his education as well—that is to say, not only what he really is by nature, but what he has appropriated from the common wealth of mankind; three-fourths of what he says does not belong to him, but has been acquired from without; so that we are often surprised to hear such a minotaur speak so humanly. And on a still further acquaintance, the brutality of which his face gave promise, will reveal itself in all its glory. Therefore a man who is gifted with a keen sense of physiognomy should pay careful attention to those verdicts prior to a further acquaintance, and therefore genuine. For the face of a man expresses exactly what he is, and if he deceives us it is not his fault but ours. On the other hand, the words of a man merely state what he thinks, more frequently only what he has learnt, or it may be merely what he pretends to think. Moreover, when we speak to him, nay, only hear others speak to him, our attention is taken away from his real physiognomy; because it is the substance, that which is given fundamentally, and we disregard it; and we only pay attention to its pathognomy, its play of feature while speaking. This, however, is so arranged that the good side is turned upwards.

When Socrates said to a youth who was introduced to him so that he might test his capabilities, "Speak so that I may see you" (taking it for granted that he did not simply mean "hearing" by "seeing"), he was right in so far as it is only in speaking that the features and especially the eyes of a man become animated, and his intellectual powers and capabilities imprint their stamp on his features: we are then in a position to estimate provisionally the degree and capacity of his intelligence; which was precisely Socrates' aim in that case. But, on the other hand, it is to be observed, firstly, that this rule does not apply to the *moral* qualities of a man, which lie deeper; and secondly, that what is gained from an *objective* point of view by the clearer development of a man's countenance while he is speaking, is again from a *subjective* point of view lost, because of the personal relation into which he immediately enters with us, occasioning a slight fascination, does not leave us unprejudiced observers, as has already been explained. Therefore, from this last standpoint it might be more correct to say: "Do not speak in order that I may see you."

For to obtain a pure and fundamental grasp of a man's physiognomy one must observe him when he is alone and left to himself. Any kind of society and conversation with another throw a reflection upon him which is not his own, mostly to his advantage; for he thereby is placed in a condition of action and reaction which exalts him. But, on the contrary, if he is alone and left to himself immersed in the depths of his own thoughts and sensations, it is only then that he is absolutely and wholly *himself*. And any one with a keen, penetrating eye for physiognomy can grasp the general character of his whole being at a glance. For on his face, regarded in and by itself, is indicated the ground tone of all his thoughts and efforts, the *arret irrevocable* of his future, and of which he is only conscious when alone.

The science of physiognomy is one of the principal means of a knowledge of mankind: arts of dissimulation do not come within the range of physiognomy, but within that of mere pathognomy and mimicry. This is precisely why I recommend the physiognomy of a man to be studied when he is alone and left to his own thoughts, and before he has been conversed with; partly because it is only then that his physiognomy can be seen purely and simply, since in conversation pathognomy immediately steps in, and he then resorts to the arts of dissimulation which he has acquired; and partly because personal intercourse, even of the slightest nature, makes us prejudiced, and in consequence impairs our judgment.

Concerning our physiognomy in general, it is still to be observed that it is much easier to discover the intellectual capacities of a man than his moral character. The intellectual capacities take a much more outward direction. They are expressed not only in the face and play of his features, but also in his walk, nay, in every movement, however slight it may be. One could perhaps discriminate from behind between a blockhead, a fool, and a man of genius. A clumsy awkwardness characterises every movement of the blockhead; folly imprints its mark on every gesture, and so do genius and a reflective nature. Hence the outcome of La Bruyere's remark: *Il n'y a rien de si delie, de si simple, et de si imperceptible ou il n'y entrent des manieres, qui nous decelent: un sot ni n'entre, ni ne sort, ni ne s'assied, ni ne se leve, ni ne se tait, ni n'est sur ses jambes, comme un homme d'esprit.* This accounts for, by the way, that instinct *stir et prompt* which, according to Helvetius, ordinary people have of recognising people of genius and of running away from them. This is to be accounted for by the fact that the larger and more developed the brain, and the thinner, in relation to it, the spine and nerves, the greater not only is the intelligence, but also at the same time the mobility and pliancy of all the limbs; because they are controlled more immediately and decisively by the brain; consequently everything depends more on a single thread, every movement of which precisely

211

expresses its purpose. The whole matter is analogous to, nay dependent on, the fact that the higher an animal stands in the scale of development, the easier can it be killed by wounding it in a single place. Take, for instance, batrachia: they are as heavy, clumsy, and slow in their movements as they are unintelligent, and at the same time extremely tenacious of life. This is explained by the fact that with a little brain they have a very thick spine and nerves. But gait and movement of the arms are for the most part functions of the brain; because the limbs receive their motion, and even the slightest modification of it, from the brain through the medium of the spinal nerves; and this is precisely why voluntary movements tire us. This feeling of fatigue, like that of pain, has its seat in the brain, and not as we suppose in the limbs, hence motion promotes sleep; on the other hand, those motions that are not excited by the brain, that is to say, the involuntary motions of organic life, of the heart and lungs, go on without causing fatigue: and as thought as well as motion is a function of the brain, the character of its activity is denoted in both, according to the nature of the individual. Stupid people move like lay figures, while every joint of intellectual people speaks for itself. Intellectual qualities are much better discerned, however, in the face than in gestures and movements, in the shape and size of the forehead, in the contraction and movement of the features, and especially in the eye; from the little, dull, sleepy-looking eye of the pig, through all gradations, to the brilliant sparkling eye of the genius. The *look of wisdom*, even of the best kind, is different from that of *genius*, since it bears the stamp of serving the will; while that of the latter is free from it. Therefore the anecdote which Squarzafichi relates in his life of Petrarch, and has taken from Joseph Brivius, a contemporary, is quite credible—namely, that when Petrarch was at the court of Visconti, and among many men and titled people, Galeazzo Visconti asked his son, who was still a boy in years and was afterwards the first Duke of Milan, to pick out *the wisest man* of those present. The boy looked at every one for a while, when he seized Petrarch's hand and led him to his father, to the great admiration of all present. For nature imprints her stamp of dignity so distinctly on the distinguished among mankind that a child can perceive it. Therefore I should advise my sagacious countrymen, if they ever again wish to trumpet a commonplace person as a genius for the period of thirty years, not to choose for that end such an inn-keeper's physiognomy as was possessed by Hegel, upon whose face nature had written in her clearest handwriting the familiar title, *commonplace person*. But what applies to intellectual qualities does not apply to the moral character of mankind; its physiognomy is much more difficult to perceive, because, being of a metaphysical nature, it lies much deeper, and although moral character is connected with the constitution and with the organism, it is not so immediately connected, however, with definite parts of its system as is intellect. Hence, while each one makes a public show of his intelligence, with which he is in general quite satisfied, and tries to display it at every opportunity, the moral qualities are seldom brought to light, nay, most people intentionally conceal them; and long practice makes them acquire great mastery in hiding them.

Meanwhile, as has been explained above, wicked thoughts and worthless endeavours gradually leave their traces on the face, and especially the eyes. Therefore, judging by physiognomy, we can easily guarantee that a man will never produce an immortal work; but not that he will never commit a great crime.

ON SUICIDE.

As far as I can see, it is only the followers of monotheistic, that is of Jewish, religions that regard suicide as a crime. This is the more striking as there is no forbiddance of it, or even positive disapproval of it, to be found either in the New Testament or the Old; so that teachers of religion have to base their disapprobation of suicide on their own philosophical grounds; these, however, are so bad that they try to compensate for the weakness of their arguments by strongly expressing their abhorrence of the act—that is to say, by abusing it. We are told that suicide is an act of the greatest cowardice, that it is only possible to a madman, and other absurdities of a similar nature; or they make use of the perfectly senseless expression that it is *"wrong,"* while it is perfectly clear that no one has such indisputable right over anything in the world as over his own person and life. Suicide, as has been said, is computed a crime, rendering inevitable—especially in vulgar, bigoted England—an ignominious burial and the confiscation of the property; this is why the jury almost always bring in the verdict of insanity. Let one's own moral feelings decide the matter for one. Compare the impression made upon one by the news that a friend has committed a crime, say a murder, an act of cruelty or deception, or theft, with the news that he has died a voluntary death. Whilst news of the first kind will incite intense indignation, the greatest displeasure, and a desire for punishment or revenge, news of the second will move us to sorrow and compassion; moreover, we will frequently have a feeling of admiration for his courage rather than one of moral disapproval, which accompanies a wicked act. Who has not had acquaintances, friends, relatives, who have voluntarily left this world?

And are we to think of them with horror as criminals? *Nego ac pernego!* I am rather of the opinion that the clergy should be challenged to state their authority for stamping—from the pulpit or in their writings—as a *crime* an act which has been committed by many people honoured and loved by us, and refusing an honourable burial to those who have of their own free will left the world. They cannot produce any kind of Biblical authority, nay, they have no philosophical arguments that are at all valid; and it is *reasons* that we want; mere empty phrases or words of abuse we cannot accept. If the criminal law forbids suicide, that is not a reason that holds good in the church; moreover, it is extremely ridiculous, for what punishment can frighten those who seek death? When a man is punished for trying to commit suicide, it is his clumsy failure that is punished.

The ancients were also very far from looking at the matter in this light. Pliny says: *"Vitam quidem non adeo expetendam censemus, ut quoque modo trahenda sit. Quisquis es talis, aeque moriere, etiam cum obscoenus vixeris, aut nefandus. Quapropter hoc primum quisque in remediis animi sui habeat: ex omnibus bonis, quae homini tribuit natura, nullum melius esse tempestiva morte: idque in ea optimum, quod illam sibi quisque praestare poterit."* He also says: *"Ne Deum quidem posse omnia. Namque nec sibi potest mortem consciscere, si velit, quod homini dedit optimum in taniis vitae poenis,"* etc.

In Massilia and on the island of Ceos a hemlock-potion was offered in public by the magistrate to those who could give valid reasons for quitting this life. And how many heroes and wise men of ancient times have not ended their lives by a voluntary death! To be sure, Aristotle says "Suicide is a wrong against the State, although not against the person;" Stobaeus, however, in his treatise on the Peripatetic ethics uses this sentence: [Greek: *pheukton de ton bion gignesthai tois men agathois en tais agan atychiais tois de kakois kai en tais agan eutychiais*]. (*Vitam autem relinquendam esse bonis in nimiis quidem miseriis pravis vero in nimium quoque secundis*) And similarly: [Greek: Dio kai gamaesein, kai paidopoiaesesthai, kai politeusesthai], etc.; [Greek: kai katholou taen aretaen aokounta kai menein en to bio, kai palin, ei deoi, pote di anankas apallagaesesthai, taphaes pronoaesanta] etc.

(Ideoque et uxorem ducturum, et liberos procreaturum, et ad civitatem accessurum, etc.; *atque omnino virtutem colendo tum vitam servaturum, tum iterum, cogente necessitate, relicturum,* etc.) And we find that suicide was actually praised by the Stoics as a noble and heroic act, this is corroborated by hundreds of passages, and especially in the works of Seneca. Further, it is well known that the Hindoos often look upon suicide as a religious act, as, for instance, the self-sacrifice of widows, throwing oneself under the wheels of the chariot of the god at Juggernaut, or giving oneself to the crocodiles in the Ganges or casting oneself in the holy tanks in the temples, and so on. It is the same on the stage—that mirror of life. For instance, in the famous Chinese play, *L'Orphelin de la Chine*,[19] almost all the noble characters end by suicide, without indicating anywhere or it striking the spectator that they were committing a crime. At bottom it is the same on our own stage; for instance, Palmira in *Mahomet*, Mortimer in *Maria Stuart*, Othello, Countess Terzky. Is Hamlet's monologue the meditation of a criminal? He merely states that considering the nature of the world, death would be certainly preferable, if we were sure that by it we should be annihilated. But *there lies the rub*! But the reasons brought to bear against suicide by the priests of monotheistic, that is of Jewish religions, and by those philosophers who adapt themselves to it, are weak sophisms easily contradicted.[20] Hume has furnished the most thorough refutation of them in his *Essay on Suicide*, which did not appear until after his death, and was immediately suppressed by the shameful bigotry and gross ecclesiastical tyranny existing in England. Hence, only a very few copies of it were sold secretly, and those at a dear price; and for this and another treatise of that great man we are indebted to a reprint published at Basle. That a purely philosophical treatise originating from one of the greatest thinkers and writers of England, which refuted with cold reason the current arguments against suicide, must steal about in that country as if it were a fraudulent piece of work until it found protection in a foreign country, is a great disgrace to the English nation. At the same time it shows what a good conscience the Church has on a question of

this kind. The only valid moral reason against suicide has been explained in my chief work. It is this: that suicide prevents the attainment of the highest moral aim, since it substitutes a real release from this world of misery for one that is merely apparent. But there is a very great difference between a mistake and a crime, and it is as a crime that the Christian clergy wish to stamp it. Christianity's inmost truth is that suffering (the Cross) is the real purpose of life; hence it condemns suicide as thwarting this end, while the ancients, from a lower point of view, approved of it, nay, honoured it. This argument against suicide is nevertheless ascetic, and only holds good from a much higher ethical standpoint than has ever been taken by moral philosophers in Europe. But if we come down from that very high standpoint, there is no longer a valid moral reason for condemning suicide. The extraordinarily active zeal with which the clergy of monotheistic religions attack suicide is not supported either by the Bible or by any valid reasons; so it looks as if their zeal must be instigated by some secret motive. May it not be that the voluntary sacrificing of one's life is a poor compliment to him who said, [Greek: panta kala lian]?[21]

In that case it would be another example of the gross optimism of these religions denouncing suicide, in order to avoid being denounced by it.

* * * * *

As a rule, it will be found that as soon as the terrors of life outweigh the terrors of death a man will put an end to his life. The resistance of the terrors of death is, however, considerable; they stand like a sentinel at the gate that leads out of life. Perhaps there is no one living who would not have already put an end to his life if this end had been something that was purely negative, a sudden cessation of existence. But there is something positive about it, namely, the destruction of the body. And this alarms a man simply because his body is the manifestation of the will to live.

Meanwhile, the fight as a rule with these sentinels is not so hard as it may appear to be from a distance; in consequence, it is true, of the antagonism between mental and physical suffering. For instance, if we suffer very great bodily pain, or if the pain lasts a long time, we become indifferent to all other troubles: our recovery is what we desire most dearly. In the same way, great mental suffering makes us insensible to bodily suffering: we despise it. Nay, if it outweighs the other, we find it a beneficial distraction, a pause in our mental suffering. And so it is that suicide becomes easy; for the bodily pain that is bound up with it loses all importance in the eyes of one who is tormented by excessive mental suffering. This is particularly obvious in the case of those who are driven to commit suicide through some purely morbid and discordant feeling. They have no feelings to overcome; they do not need to rush at it, but as soon as the keeper who looks after them leaves them for two minutes they quickly put an end to their life.

* * * * *

When in some horrid and frightful dream we reach the highest pitch of terror, it awakens us, scattering all the monsters of the night. The same thing happens in the dream of life, when the greatest degree of terror compels us to break it off.

* * * * *

Suicide may also be looked upon as an experiment, as a question which man puts to Nature and compels her to answer. It asks, what change a man's existence and knowledge of things experience through death? It is an awkward experiment to make; for it destroys the very consciousness that awaits the answer.

FOOTNOTES:

19   Translated by St. Julien, 1834.
20   See my treatise on the *Foundation of Morals*, Sec. 5.
21   Bd. I. p. 69.

Lightning Source UK Ltd.
Milton Keynes UK
UKHW022328060223
416579UK00001B/193